A GUIDE TO FOSSIL HUNTING ON THE EAST DORSET COAST

Steve Snowball
&
Craig Chivers

With prehistoric life and scene reconstructions by
Andreas Kurpisz

Written by Steve Snowball & Craig Chivers.

All photographs by the authors, except where stated.

©2021, Siri Scientific Press. All rights reserved. No parts of this publication may be reproduced, stored in a retrieval system or transmitted, in any form or by any means, electronic, mechanical, photocopying, recording or otherwise, without the prior written permission of the publisher. This does not cover photographs and other illustrations provided by third parties, who retain copyright of their images; reproduction permission for these images must be sought from the copyright holders.

ISBN: 978-1-8381528-2-6

Published by Siri Scientific Press, Manchester, UK This and related titles are available directly from the publisher at **http://www.siriscientificpress.co.uk**

The authors have made every effort to contact copyright holders of material reproduced in this book and are grateful for permissions granted. If any have been inadvertently overlooked, we apologise for our oversight, which we hope will be accepted in good faith.

Steve Snowball & Craig Chivers have asserted their right under the Copyright, Design and Patents Act 1988 to be identified as the authors of this work. All author profits from this publication will support the work of the Etches Collection: Museum of Jurassic Marine Life; The Kimmeridge Trust Charity No. 1106638.

Cover and book design: Steve Snowball & Craig Chivers.

Front cover artwork: Scene with *Steneosaurus* and *Rhomaleosaurus* by Andreas Kurpisz.

DISCLAIMER: This book includes reference to fossil collecting localities but is not an extensive manual for health and safety when visiting such sites. Furthermore, because potential hazards may change over time, prior to undertaking any fossil collecting activities, you need to make yourself aware of any RISKS, DANGERS, HAZARDS and LEGAL IMPLICATIONS associated with visiting and collecting fossils at any particular site. The publisher, authors or any associated parties cannot be held responsible for your failure to do so, nor any consequences thereof. Enjoy your fossil collecting safely and responsibly.

Printed and bound in the UK

Fossil 'Forest': algal burrs' which surrounded the bases of the trees in a Jurassic cypress forest, Lulworth Cove.

A GUIDE TO FOSSIL COLLECTING ON THE EAST DORSET COAST

Steve Snowball & Craig Chivers

With prehistoric life & scene reconstructions by
Andreas Kurpisz

Chalk cliffs below Bat's Head, near Ringstead Bay.

ACKNOWLEDGEMENTS

The authors thank and acknowledge the contributions of the following people and organisations, for their help and advice in the preparation of this book:

Andreas Kurpisz, once again, for his valued and inspiring artistic contributions of reconstructed scenes of Jurassic and Cretaceous life.

Vivien Field for her much-welcomed contribution in supplying many of the photos used within these pages.

The following people and organisations have also helped to contribute towards the writing of this book. We gratefully thank:

Prof. David Bridgland, Dr. Robert B. Chandler, Carla Crook, Alister Cruickshanks, Ian Cruickshanks, Martin Curtis, Dr. Steve Etches, Bill Fagg, Lizzie Hingley, Anastasia Jeune, Julian & Vicki Sawyer, Roy Shepherd, Dr. Steven Sweetman, Hashimoto Tadanori, Andy Temple, Dr. Ian West, Dr. Mark Witton.

The authors are especially indebted to the Etches Collection Museum of Jurassic Marine Life for their kind support and permissions for the use of many images in this book.

The Smedmore Estate
The Geologists' Association
Charmouth Heritage Coast Centre
The Jurassic Coast Trust
Purbeck Footprints
Swanage Museum

Dr. David Penney, owner and publisher of Siri Scientific Press, for his continued support and help in the production of this publication and others in the series.

Finally, thanks to family, friends and all others who have helped, encouraged and supported us during the process of writing this book.

Plastron (the underside) of a marine turtle
from the Lower Cretaceous,
Purbeck Beds, Swanage.
Image used with kind permission of Swanage Museum.

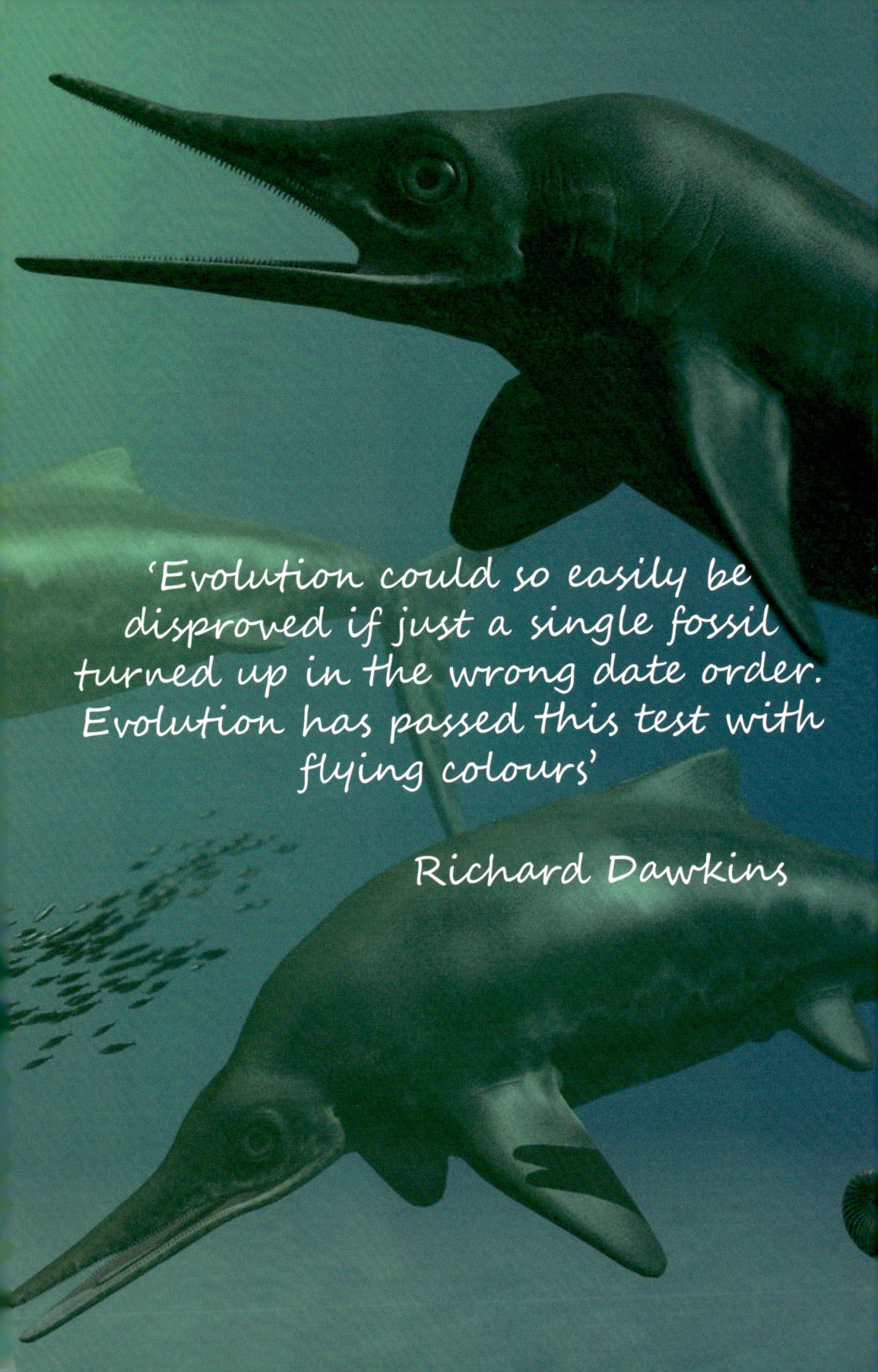

'Evolution could so easily be disproved if just a single fossil turned up in the wrong date order. Evolution has passed this test with flying colours'

Richard Dawkins

A group of ichthyosaurs and a *Dakosaurus maximus* in the Kimmeridgian sea of the late Jurassic. Artwork by Andreas Kurpisz.

CONTENTS

ACKNOWLEDGEMENTS — 6

FOREWORD BY THE AUTHORS — 14

1. INTRODUCING THE JURASSIC COAST: — 16
A World Heritage Site

2. DORSET FOSSIL COLLECTING CODE OF CONDUCT — 22

3. THE EAST DORSET SUCCESSION: — 38
The geology & landscape of East Dorset

4. MAKING SENSE OF THE FOSSIL RECORD — 51

5. FOSSIL COLLECTING EXCURSIONS IN EAST DORSET: — 76

   i. EXCURSIONS NEAR DURDLE DOOR — 78
      Bat's Head to St. Oswald's Bay, Dungy Head & Lulworth Cove

   ii. EXCURSIONS NEAR WORBARROW BAY — 94
      Mupe Bay, Worbarrow Bay, Brandy Bay & Gad Cliff

   iii. EXCURSIONS NEAR KIMMERIDGE BAY — 104
      Broad Bench towards Chapman's Pool

iv. EXCURSIONS NEAR ST. ALDHELM'S HEAD   124
    St. Aldhelm's Head & the Purbeck Quarries

v. EXCURSIONS NEAR DURLSTON BAY   144
    Durlston Bay, Peveril Point & Swanage

vi. EXCURSIONS NEAR HENGISTBURY HEAD   178
    Studland Bay & Hengistbury Head

6. ABOUT THE AUTHORS   186

7. BIBLIOGRAPHY   188

8. INDEX OF FOSSIL PHOTOGRAPHS   190

Late Jurassic scene with *Dacentrurus armatus* taking a fatal step into a fast-flowing river, to be swept out to sea. Remains of this early British stegosaur have been found in the Kimmeridge Clay Formation. Artwork by Andreas Kurpisz.

FOREWORD BY THE AUTHORS

This book concludes our fossil collecting journey along the beautiful Jurassic Coast of Dorset, which began in our first book in the series, '*A Guide to Fossil Collecting on the West Dorset Coast*' (2018). Our second book, '*A Guide to Fossil Collecting on the South Dorset Coast*' (2020) ended at the majestic Chalk headland of White Nothe at the eastern end of Ringstead Bay, which is very much where this book begins. From here we travel eastwards, sometimes to the wildest and remotest parts of the Dorset coast and where many sections are totally inaccessible, due to the nature of the terrain. In these areas, only the convenience of the Dorset Coastal path allows glimpses of the coast and the tantalising fossil-bearing rocks far below.

As we know, fossil collecting can involve risk. However, at all times we advocate safe and responsible collecting practices but we need to emphasise that this particular section of the Dorset coast can pose significant dangers, especially with regard to the Kimmeridge cliffs. The main danger is the risk of rock falls and the second hazard is the very real risk being cut off by the tide. At Kimmeridge Bay and eastwards, the cliffs are vertical and high and subject to intense erosion at their base. Substantial, frequent rock falls, at high velocity, can occur without any warning and traversing the section will certainly require the prudent use of a hard hat at all times. You will also require a low tide to travel along the cliffs to the east and the danger of being cut off there is very real, unless proper precautions are undertaken.

Always check and double check tide times and be aware of double tides. In any case, the actual collection of fossil material at Kimmeridge Bay and to the east is prohibited, unless prior permission is granted by the Smedmore Estate. The landowners own a large section of coast around Kimmeridge and are rightfully concerned about collection methods often employed. Under no circumstances should visitors use geological hammers to extract fossils from the cliffs or bedrock. This is an infringement of the Dorset Fossil Collecting Code, goes against the wishes of the landowner and the area has SSSI status (Site of Special Scientific Interest) awarded to it. A better alternative is to view a spectacular range of fossils found from this vicinity, now displayed at the Etches Museum at Kimmeridge Bay, which houses specimens that most of us can only dream of finding and where they can be seen and studied, both in context and in safety.

The mudstones and shale of the Kimmeridge Clay extend from Brandy Bay to St. Aldhelm's Head, in other words a large part of the east Dorset coastline. Debris falls every day and in certain weather conditions these falls can be substantial. It is a dangerous place, unless it is well understood. Always keep well away from the base of the cliffs. The higher cliffs to the east are even more dangerous than those at Kimmeridge Bay and you will certainly need a low tide to travel far.

It might be that geologists or fossil collectors find the above safety advice a bit too daunting, but other locations described in this book might be considered to be less hazardous and less restrictive. Again though, we advocate the need for a well-planned visit, which takes the local terrain and tide times into consideration. As is the case with the rest of the Dorset coast, the section to the east of the county also has SSSI status and collectors should be aware that the collection of any fossil material should be from loose material (*ex-situ*) only. There should be strictly no hammering or digging into the bedrock or cliffs under any circumstances.

The authors take no responsibility for any activities of field parties or individuals going to the coast for their own purposes or objectives. As at other geological sites, a risk is present and the possibility of an accident, although a rare occurrence, cannot be eliminated. East of Kimmeridge Bay, the Army might well be active on the Lulworth ranges, so accessibility is often very restricted during the published firing times. Again, it will be prudent to check before setting out and to avoid disappointment. Furthermore, the Enscombe Estate, the land owners at Chapman's Pool, do not wish for fossil collecting on their land, other than by authorised persons.

Of course, taking an interest in the basic geology of the area certainly aids with the potential collection of fossils and an understanding of rock types can help locate the best sites and aid with the identification of your finds. The geology of Dorset is often complex but we hope this guide can help with overcoming some of the hurdles of finding suitable fossil-bearing locations. The east coast of Dorset is unsurpassed in the landforms which dominate the coastal scenery and even if accessibility to the beaches proves difficult, the South West Coastal Path is certainly a means by which the dramatic landscape can be taken in and enjoyed.

Steve Snowball & Craig Chivers
2021

Fossil 'Forest'. Lulworth Cove.

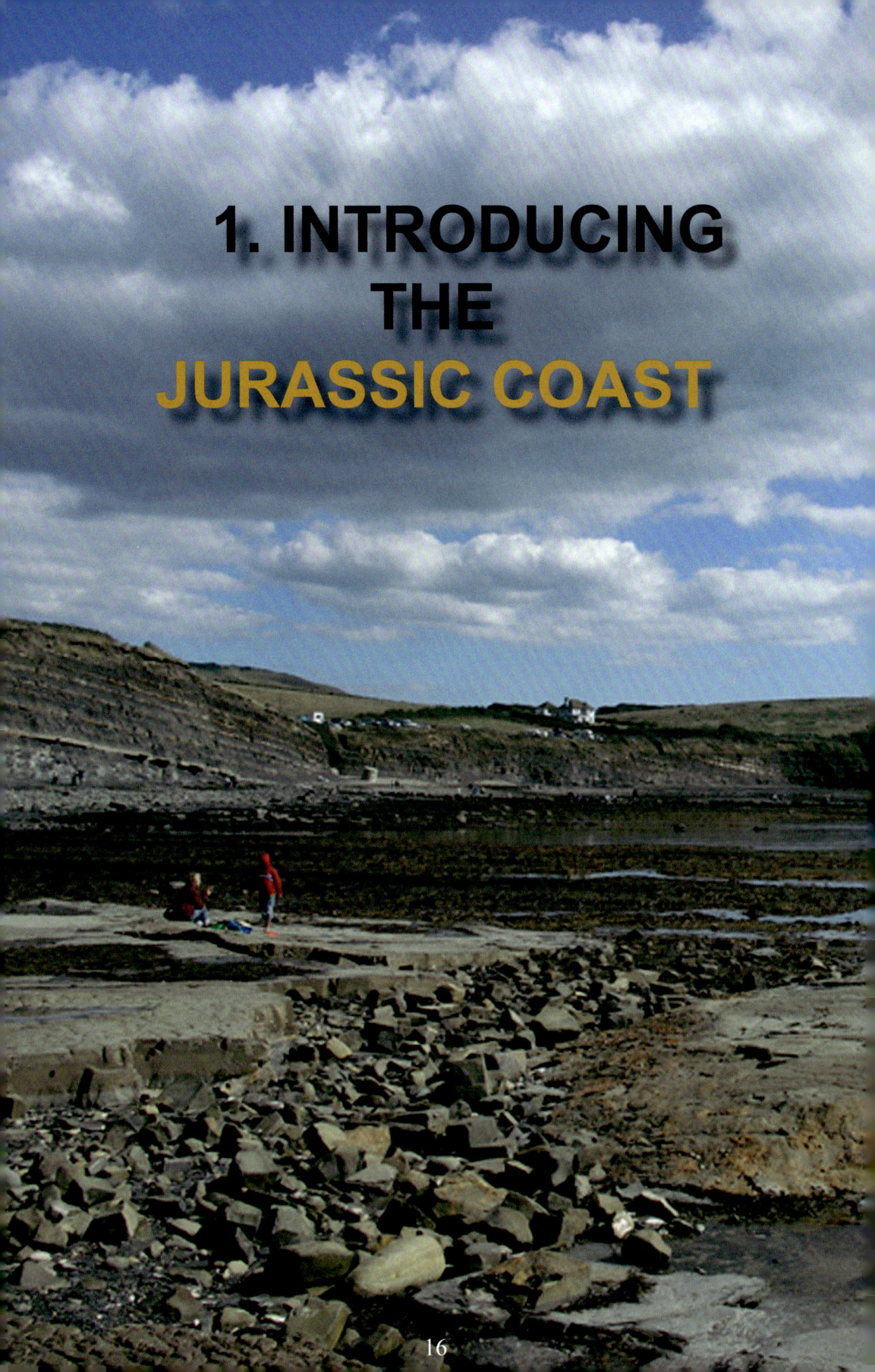

1. INTRODUCING THE JURASSIC COAST

Kimmeridge Bay west to Clavell Tower.

A WORLD HERITAGE SITE

St. Oswald's Bay, near Lulworth.

"The Dorset and East Devon Coast has an outstanding combination of globally significant geological and geomorphological features.
Along 155 km of largely undeveloped coast the Site's geology displays approximately 185 million years of the Earth's history, including a number of internationally important fossil localities".

United Nations Educational, Scientific & Cultural Organization

Stair Hole, Lulworth.
The Jurassic Coast of Dorset and East Devon is a UNESCO World Heritage Site and was designated as such in 2001, due to its outstanding rocks, fossils, landforms and landscape which encapsulate 185 million years of this planet's prehistoric past.

Much has been written about the Jurassic Coast and rightly so. The Jurassic Coast of Devon and Dorset, with its unique, natural landforms and breath-taking scenery, has led to its international recognition and the designation as a UNESCO World Heritage Site in 2001.

The coast of East Dorset is just a part of this magnificent region of Britain. It is an area of significant interest and importance to geologists and fossil collectors, both amateurs and professionals alike. It is also an area of outstanding and often contrasting scenery; from the majestic Chalk cliffs of White Nothe or the hard limestone outcrops of Portland Stone and Purbeck limestones around Lulworth, to the shale and clay cliffs at Kimmeridge Bay. Whilst not as celebrated as the wild, rugged western section of the county's coastline, nor as famous as Lyme Regis and Charmouth, the fossils found along the eastern coast have nonetheless attracted scientific attention for well over two hundred years.

This area is part of the Jurassic Coast of Devon and Dorset, which stretches from Orcombe Point, near Exmouth to Old Harry Rocks, near Swanage. The layers of rocks which run its length provides a continuous exposure of rocks of Mesozoic age and provide a 'walk in time,' through 185 million years of Earth history in just 160 kilometres. There is no other rock sequence on this planet that is exposed in this way. It is also one of the few places on this planet that holds a unique record of 185 million years of Earth's pre-history in the rocks found along its length. The World Heritage Site status bestowed upon it will ensure that the region and indeed the nation, benefit from this fascinating and ever-evolving coastline for generations to come.

Mupe Rocks, from the coastal path, looking east.

2. DORSET FOSSIL COLLECTING CODE OF CONDUCT

Kimmeridge Clay cliffs, east of Kimmeridge Bay.

In our previous books, *A Guide to Fossil Collecting on the West Dorset Coast* and *A Guide to Fossil Collecting on the South Dorset Coast,* the West Dorset Fossil Collecting Code was printed in its entirety. It is a hugely important document and forms *the* guiding reference for *all* collectors in the area. In East Dorset, a guiding code is extremely relevant and a differing set of protocols needs to be observed, especially on land under the ownership of the Smedmore Estate in the Kimmeridge area.

A Working Group of landowners, conservation organisations, museum curators and local fossil collectors was established in order to address growing conflicts of interest with regard to fossil collecting along the West Dorset coast. The Group recognised the essential need for fossil collecting to continue in such a way as to satisfy all those with an interest in our fossil heritage. In the eastern part of the county, no such working group has been established but it is true to state that many organisations agree with the premise of a responsible approach to collecting, especially on private land. It may surprise many to know that whilst fossil collecting generally presents no threat to landowners, land owned or managed by the National Trust, the Smedmore or Edgecombe Estates, the MOD and others, should not be taken as a public right of way, even along the coast. Ownership of fossils of importance collected on such land is as much the property of the landowner as the person who found them.

Land is also often subject to local bye-laws. Care and consideration for the fragility of wildlife in this area is of utmost importance. Also, all sections along the Dorset coast have SSSI (Site of Special Scientific Interest) status, which forbids digging or hammering into cliffs and *in situ* rocks.

However, on the rapidly eroding Dorset coast, fossil collecting is often essential if specimens, some of which may be of great scientific value, are to be saved from damage or destruction by the sea. Collecting offers an opportunity for people to learn about the prehistoric past and to contribute to our understanding through the discovery of new finds or the development of scientific study. Most collectors, both amateur and professional, have a deep-seated interest in palaeontology and a wish to contribute to the development of the science.

As a general rule, **Landowners** own the fossils on or under their land. The **National Trust** is a principal landowner; a registered charity charged with preserving places of Historic Interest or Natural Beauty for the nation to enjoy. **Smedmore Estate** owns a large part of the land and coast around Kimmeridge. **Natural England** is the Government's statutory advisor on conservation, including the earth sciences. It designates National Nature Reserves and Sites of Special Scientific Interest and promotes sustainable management of these sites.
Museum curators and researchers are keen to secure key scientifically important specimens for recognised collections as part of the nation's heritage and to provide a collection upon which scientific research can be based. They seek to ensure that the maximum associated scientific data are gathered when specimens are collected. Some researchers require access to strata and specimens *in situ* in order to undertake their work.

Opposite page: *Lepidotes* sp. Early Cretaceous Fish, Intermarine Beds, Purbeck Group, Image used with kind permission of Swanage Museum.

Objectives of a Code

The interests of all those involved with fossil collecting on the Dorset Coast need not be mutually exclusive. Indeed, many interest groups can assist each other so long as each party is aware of, and accepts the interest of the other. A voluntary fossil collecting Code of Conduct attempts to balance those interests. Whereby the primary objectives of the code are to:

1. promote responsible and safe fossil collecting,
2. restrict the excessive digging or 'prospecting' for fossils along fossil rich strata,
3. clarify ownership of the fossils,
4. promote better communication between all those with an interest in fossils,
5. promote the acquisition of key scientifically important fossils by recognised museum collections.

Health & Safety

The following is a general list of practical advice aimed at all types of collector, including professionals and amateurs, educational/academic visitors and the general public including holiday makers and local people.

• Always consult tide tables before collecting. It is advisable that you go collecting on a falling tide.

• Always advise someone of where you are going and at what time you can be expected to return.

• Be vigilant and exercise common sense in the vicinity of any cliffs. Cliff falls tend to occur suddenly and without warning. Avoid cliff bases.

• Avoid walking on, and keep clear of, visibly moving rock falls and mudflows. Note particularly that the seaward edges of mudflows may be covered by shingle and can be particularly treacherous.

• If you are using a hammer or other tools, it is advisable to wear safety goggles.

• Exercise common sense when considering what clothes and safety items to wear and take with you.

• Collectors should not descend the cliffs using ropes to get to a particular level under any circumstances.

Opposite page: *Acanthoceras jukesbrownei* ammonite from the Grey Chalk Subgroup, Swanage Bay. Found and photographed by Viv Field.

For professional and experienced amateurs collecting from cliffs and the foreshore, the Code provides as follows:

- There should be no digging *in situ* in the cliffs. SSSI status applies.
- Collectors should adopt a common sense approach to their activities and not expose themselves to excessive risks. Before visiting a location it is highly advisable to research the potential dangers and necessary precautions. They should cease immediately on becoming aware that their activities present a risk to a third party.
- Collectors should take particular care in connection with the following features:

1. Unstable cliffs, especially in areas where recent cliff falls have occurred or are ongoing. This particularly applies to the section from Kimmeridge Bay to Durlston Bay. Along this stretch of the coast, rockfalls are commonplace and the cliffs prone to collapse.

2. Tides, rough seas and poor weather conditions. Rising tidal water often isolates sections of the coast and may block your path.

Cliff excavations: Collectors wishing to extract fossils from the cliffs should use the following procedure:
- Obtain the landowner's permission *before* taking *any* action to excavate any part of the find.
- Prepare a Risk Assessment (RA) for the excavation to identify the hazards that may arise in the course of the excavation, and the precautions that should be adopted, to protect the collector and others in the vicinity. This should then be discussed with the landowner.

Items that the RA is likely to cover are as follows:
To cordon off the area of working.
To ensure, as far as practicable, the stability of the surrounding area during the excavation. Effective communication among all parties involved in the excavation (including the landowner), and a procedure for dealing with accidents or problems arising from the work.
- To ensure, as far as practicable, that the site is safe when left unattended, and that appropriate signage, etc., is in place. This list is by no means exhaustive and collectors should satisfy themselves that all risks have been assessed.

3. Keep the landowner informed of progress with the excavation, and advise when completed.

4. In the event of a fossil being located which is at *immediate* risk of being lost or damaged, the collector may proceed with the excavation provided that he gives full consideration to the risks and takes appropriate action to alleviate them, and is satisfied that the work will present no risk to any third parties. The collector should notify the landowner at the next available opportunity.

5. In Kimmeridge Bay, or adjacent coastal sections, if you think you have found something of significance, contact the Etches Collection Museum on 01929 270000.

Opposite page: Kimmeridgian cuttlefish, Kimmeridge Clay Formation, Kimmeridge, Dorset. Image used with kind permission of the ©Etches Collection.

Macromesodon daviesi
Early Cretaceous (Berriasian) pycnodont fish,
Intermarine Beds of the Stair Hole Member,
Durlston Formation, Purbeck Group,
Keates Quarry, Worth Matravers. Found by Kevin Keates Snr.
Prepared by Craig Chivers.

Key Scientifically Important Fossils Recording Scheme

There are two categories of fossils recognised within the Recording Scheme: **Category I, Key Scientifically Important Fossils,** and **Category II, Fossils of some (but not key) importance.**

Category I Fossils include new species or those specimens which may represent new species, fossils which are extremely rare and fossils that exhibit exceptional preservation. This includes vertebrates such as terrestrial and marine reptiles, pterosaurs and fish, partial or complete.

Category II Fossils include vertebrates, especially where the horizon of origin can be identified. Nautiloids and certain ammonites together with unusual assemblages of fossils are also included.

To comply with the Code, all **Category I** fossils are to be reported and certain restrictions may apply to their disposal (see end of this section). To comply with the Code, it is not obligatory to record **Category II** fossils although it is strongly recommended. No restrictions apply to the disposal of Category II fossils.

1. All Category I records should include an identification of the specimen (if known), a photograph, the exact location of the find together with the scientific horizon (if known), the date of the find and any other related observations. The name of the collector will be kept with the record but may not be available directly within public records depending upon the wishes of the individual.

2. In the case of fossils found in the Kimmeridge Clay, the Etches Museum will photograph the specimen and the record will be kept in paper form and on an internet site. They will, as and where necessary, act as an intermediary between collectors and other interested parties.

3. Where a specimen is being recovered over a protracted period, it should still be recorded but the exact location of the site may be withheld in order to protect the finder's interest until the specimen has been fully recovered.

4. The preparation of Category I specimens should only proceed after consultation with appropriate academics or museum curators unless preparation is clearly straight forward, or work needs to be carried out urgently.

5. Under the Code, collectors who intend to sell their Category I specimens will offer them to registered museums for a period of six months. If no purchase has been agreed by this time, the collector will be free to offer the specimen elsewhere. The recording scheme should be updated accordingly.

6. Those individuals with private collections that contain Category I specimens are encouraged to make provision for the ultimate placement of such specimens within registered museums.

7. The scheme offers a channel of communication for curators and researchers to convey their interests to collectors.

Fossil Ownership

Landowners generally wish to make clear their ownership of these fossils but they are often willing to see ownership transferred to those collectors who have followed the Fossil Collecting Code of Conduct and have recorded their key scientifically important fossils.

Maps of land ownership will be provided by the National Trust, Smedmore Estate and the MoD upon request.

Please note: Those collectors who do not follow this voluntary code, particularly by digging or prospecting *in situ* in the cliffs, or failing to record Category I fossils, may be regarded as stealing the fossils, and appropriate legal action may be taken against them.

Key Scientifically Important Fossils

The Jurassic and Cretaceous rocks exposed on the Dorset coast contain abundant and extremely diverse fossils. Therefore, the following lists aim to provide general guidance only and are not to be regarded as fully comprehensive. Wherever there is doubt about the scientific importance of any fossil finds, collectors are recommended to contact the relevant fossil group specialist(s) for assistance.

Category I fossils

1. Fossils which *certainly* represent new species. These can belong to any taxonomic group – vertebrate, invertebrate or plant.

2. Fossils that belong to any group that are *thought* to represent new species.

3. Fossils that are extremely rare. Although not necessarily new species, they are nevertheless clearly of great scientific importance. Examples include dinosaurs, significant reptiles, pterosaurs, sharks and rays, (near) complete insects and arthropods (crustaceans, crabs), early mammals, recognisable leaf fronds and plant cones, etc.

4. Fossils which exhibit exceptional preservation. For example, ichthyosaurs (or other vertebrates) showing skin texture, uncrushed skulls which could provide data on brain size or other physiological aspects, etc.. Among invertebrates, fossil cephalopods (cuttlefish, squids, ammonites or belemnites) showing traces of gill structures, arms and hooks, etc. are of key scientific importance.

Opposite page: Footprint of a probable iguanodont dinosaur. Haysoms Quarry, Swanage, Stair Hole Member, Durlston Formation, Purbeck Group. Found by Lizzie Hingley.

Category II Fossils
Reptiles: ichthyosaurs, pliosaurs, plesiosaurs, crocodiles, turtles etc.
Fish: including sharks, rays, bony fish, etc.

Fossil remains, especially fragmentary, isolated, bones or scales, etc., may be relatively common in some beds. The stratigraphic range of many forms is poorly known and any data may be important to relevant specialists. It is recommended therefore that collectors do record significant, recognisable finds if found *in situ* or where the stratigraphic horizon can be identified satisfactorily.

Arthropods: insects
Many insect remains are indistinctly preserved, but given their scarcity, any recognisable forms are worthy of recording.

Molluscs: belemnites
Extremely common fossils especially in the form of isolated belemnite guards. It is not anticipated that these would be recorded, unless a particular bedding-plane concentration or similar fauna was collected.

Molluscs: ammonites
One of the most common and characteristic fossils from the Dorset coast, occurring throughout the section. Many of the usual taxa are abundant and comprise the 'bread and butter' specimens for commercial, amateur and tourist collectors. It is not anticipated that these forms would be recorded, although any unusual species or particularly large/mature shells showing apertural details etc. are worthy of inclusion in the database.

Molluscs: nautiloids
A neglected group of fossils, occurring throughout much of the succession and rarely collected commercially. It is not expected that these would be recorded, though exceptional specimens (e.g., bedding-plane assemblages or others yielding palaeoecological data) are worth considering for inclusion in the database.

Molluscs: bivalves
An abundant group of fossils, occurring throughout much of the succession and rarely collected commercially. It is not expected that these would be recorded or reported.

Brachiopods
An abundant group of fossils, occurring throughout much of the succession and rarely collected commercially. It is not expected that these would be recorded or reported.

Echinoderms: crinoids & starfish
A group of considerable interest to collectors. There are many specimens of these in public collections and it is not anticipated that specimens would normally be reported.

Opposite page: Compacted ammonite. Kimmeridge Clay Formation, Brandy Bay, Kimmeridge, Dorset.

Callipurbeckia minor (formerly *Lepidotes minor*)
Lower Cretaceous, Intermarine Beds, Stair Hole Member,
Durlston Formation of the Purbeck Group, Swanage.
Prepared by Craig Chivers.

The authors hope that collectors will familiarise themselves with the Code. Category I and Category II fossils of importance found in the Kimmeridge Clay sections of the Kimmeridge shoreline, should be reported to:
Steve Etches, c/o The Etches Collection, Museum of Jurassic Marine Life, Kimmeridge, BH20 5PE or via the website at https://www.theetchescollection.org/contact

Any important fossils found from this area can be donated to The Etches Collection, as part of the museum's acquisition policy and would immediately be recorded and added to the Fossil online database (accessible via the Website under the 'About' section) at https://database.theetchescollection.org/collections

Further advice about fossil collecting on the Dorset coast can also be obtained from Sam Scriven, Programme Manager & Conservation, Jurassic Coast Trust, Mountfield, Bridport, DT6 3JP Tel: 01308 80700 ext. 204. Email: sam.scriven@jurassiccoast.org

3. THE EAST DORSET SUCCESSION

The geology & landscape of East Dorset

View across the Purbeck Hills, near Tyneham Gap.

GEOLOGY & LANDSCAPE: AN OVERVIEW

"Over the last two hundred years the study of the rocks of Dorset has made a fundamental contribution to the science of geology. The best exposures of the rocks of Dorset are inevitably along its coastline, but the contribution of inland Dorset to our current understanding of geology and landscape development cannot be entirely ignored."

John Chaffey – The land on which we live – The Geology of Dorset.

The nature of the Dorset landscape is heavily influenced by the underlying geology. In the east of the county, where the rocks are soft, the landscape is gentle and rolling, whereas in the west (described in our earlier volume) they are generally harder and hence more able to withstand erosion and the landscape takes on a more undulating form. The Isle of Purbeck is renowned for the variety and structural clarity of its rocks and landforms. A high chalk ridge, which once linked Purbeck to the Isle of Wight, separates the heathlands of the Poole Basin from the secluded clay valley of the River Corfe to the South. A further ridge and distinctive plateau, this line of limestone, separates this valley from the sea.

The Kimmeridge Clay Formation is no stranger to Dorset. It is found in the southern section of the county and makes a significant contribution to the low-lying valleys and coastal sections around Weymouth. In the east of the county, it is the dominant rock formation from Brandy Bay to St. Aldhelm's Head.

Kimmeridge Bay from the east.

As we travel east from Ringstead Bay, it becomes noticeable that the rocks belonging to those of the Jurassic System (and which are highly evident in the western and central regions of the county), were uplifted. The early Cretaceous rocks, displayed in Purbeck and Wealden strata, were laid down, not in the deep waters of late Jurassic times but in shallow coastal lagoons or by rivers in flood plains, which were the main sites of deposition. Sea levels were relatively low during this time. Thus, the early Cretaceous strata includes shallow lagoonal deposits of the Purbeck Formation, similar to those found at the head of the Persian Gulf today.

Our knowledge of the climate at this time is partly based on the growth rings on the fossil trees found in the Lower Purbeck rocks. Winter rains were followed by summer drought and it has been suggested that the climate was very similar to parts of North Africa or southern Iraq, where fresh water marshes exist. The environment of Purbeckian Dorset supported all forms of life but especially *Viviparus* molluscs, whose huge numbers help to make up the depositional limestones, for which the Purbeck region is so famed.

The collision of continental plates around 385–285 Ma, known as the Variscan Oregeny, produced a variety of east to west trending structures, including folds and faults, which were thrown up right across southern Britain. This ancient structure greatly influences the Dorset landscapes and over which the sediments from millions of years of geological time have been deposited, further uplifted and eroded to create the Jurassic Coast. Plate tectonics applied stresses and strains to the very fabric of southern Dorset, where strata have been further folded and faulted and where the results of these dramatic episodes can be seen along the coast.

Stair Hole, Lulworth Cove.

Across the Isle of Purbeck, from Durdle Door to Ballard Down, the rocks have been heaved up by these enormous Earth movements into a huge 'kink,' known as the Purbeck monocline. Either side of the fold the rocks are virtually horizontal but within it, they are vertical. Where the Chalk is caught up in this fold, it forms the ridgeway extending from Lulworth to Ballard Down and out across to The Needles, on the Isle of Wight.

The Purbeck Group is renowned for its Purbeck limestones; a durable building stone, displaying a range of colours ranging from off-whites, buffs and browns to dark greens and blue. Numerous beds are highly fossiliferous and can be highly packed with shells and benthic fauna.

The Middle Purbeck strata begins in the Upper Jurassic period, but from the Cinder Bed upwards is thought to have entered the Lower Cretaceous Period, which began approximately 136 mya, although the boundary for this evolution has been the subject of much debate. The Cinder Bed, a quarryman's term, constitutes a very dense layer of oysters at the bottom of the Middle Purbeck. Within this Middle Bed are found the Lower and Upper Building Stones, directly below and above the Cinder Bed respectively. Just beyond the Middle is the Upper Purbeck which contains the Purbeck 'Marble' along with Broken Shell Limestone.

Purbeck limestone in the cliff face at Lulworth Cove.

The tyrannosaurid *Juratyrant langhami* in the salt lake section of what will become the Purbeck coast, with the pterosaur *Aurorazhdarcho primordius* in the background, flying from the coast of Spain. Artwork by Andreas Kurpisz.

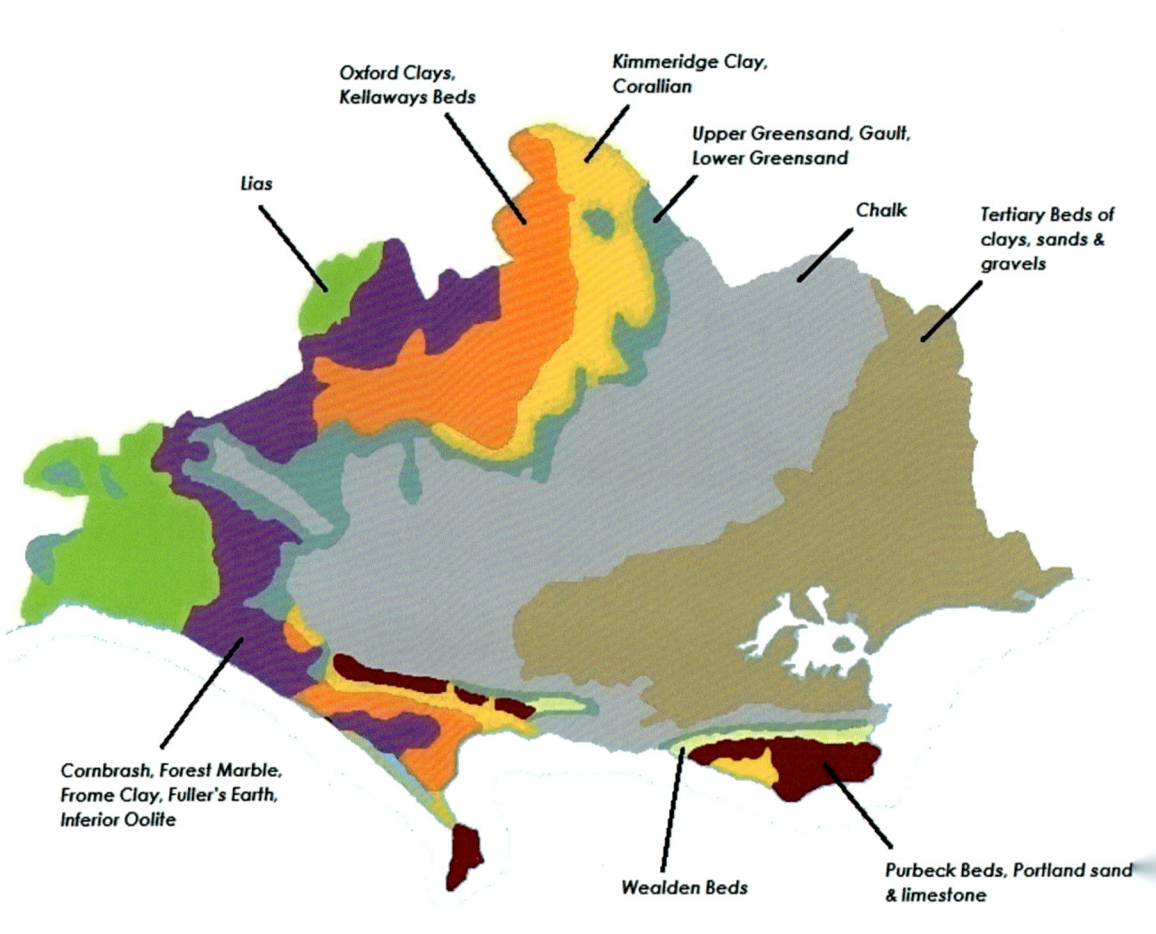

Simplified geological map of Dorset.

A reconstructed landscape during Purbeckian times, with coastal lagoons and saline forests. Artwork by Andreas Kurpisz.

Later in Wealden times, Dorset received the deposits of a huge river, which flowed into the area from the west. The clastic sediments of the Wealden beds are best seen in the main outcrop in the Isle of Purbeck and are represented by sands, sandstones, clays and grits and floor valleys between Swanage Bay and Durdle Door. They are well exposed in the cliffs of Swanage Bay, Mupe Bay, Stair Hole, Dungy Head and Durdle Door and underlie the Vale of Purbeck, westwards to Worbarrow Bay. Here, they form characteristic red, brown and yellow mottled mudstone cliffs, with thin ironstones and plant debris (a common fossil, along with charcoal fragments, suggesting forest fires were prevalent). Harder sandstone bands also form distinct ridges in the Wealden beds, such as Windmill Knap west of Swanage, and Corfe Common.

Fossils are uncommon in the Wessex Formation of Dorset and the absence of extensive foreshore sections means that the strata are not on par with exposures found on the Isle of Wight. The best section is undoubtedly at Swanage Bay, where plant debris beds, fish, crocodile and occasional dinosaur taxa have been found.

Worbarrow Bay, where Wealden clays, sands and grits form a broad valley between Swanage and Tyneham.

Following the clastic sediments of fluvial origin during Wealden times, the major unit at the top of the Cretaceous is the Chalk. In these later Cretaceous times, sea levels began to rise again, as the North Atlantic opened and Britain drifted further to the north: This was a period of Chalk deposition, where great thicknesses of Chalk were laid down in a very clear and clean sea. Sea levels rose to some 200 metres higher than today, and the Chalk was deposited over much of the present day position of the British Isles. Unsurprisingly, the nature of the Dorset landscape is heavily influenced by the band of Cretaceous Chalk, which runs from the south west to the north east of the county and forms part of the Southern England Chalk Formation.

The central Dorset section of the chalk formation forms the Dorset Downs and the north eastern section, which runs into Wiltshire, is Cranborne Chase. The two are separated by the Stour Valley, which cuts through the hills at Blandford Forum. In the Isle of Purbeck is another, smaller, ridge of the chalk formation known as the Purbeck Hills.

The Dorset Downs: View of Kingston Down promontory from West Hill, Isle of Purbeck.
The ridge is of Portland Sands and Stone.

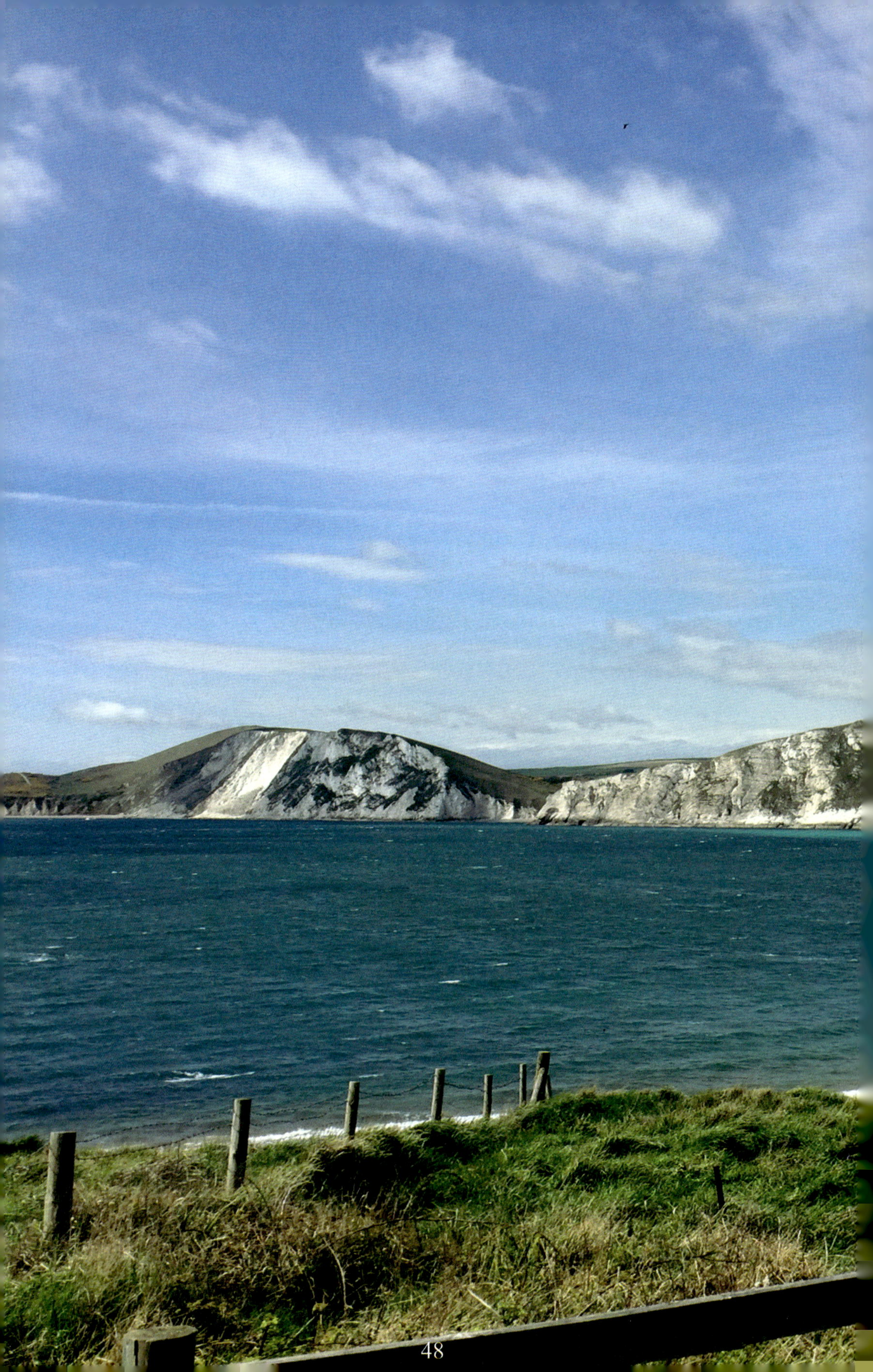

The Chalk coastline at Worbarrow Bay, extending from Mupe Bay in the west.

Sixty million years ago further uplift occurred and erosion of the Chalk began, much of it being removed completely from western Britain. After 10 million years the sea level began to rise again and pebbly beds, sands and clays were deposited, to be followed by more deposition by rivers and in lagoons and estuaries.

Essentially, the rocks exposed along the Dorset Coast decrease in age from west to east. Along the coast of the World Heritage Site, which extends from Lyme Regis to Studland, Jurassic rocks first make their appearance at the Devon/Dorset boundary in the west, and continue, with a small number of breaks to Durlston Head at Swanage. Cretaceous rocks, which overlie the Jurassic sequence of rocks, first appear in Dorset in the upper part of the cliff at the Devon/Dorset boundary. Eastwards they form the summits of the cliffs as far east as Thorncombe Beacon. They reappear in the Chalk cliffs of White Nothe to the east of Weymouth, and continue to form important parts of the coast as far east as Worbarrow Bay, cut in the Wealden Beds.

Cretaceous gives way to Jurassic rocks again as far as Durlston Head. They reappear at Durlston Bay as the Purbeck Beds. The Wealden Beds appear again in Swanage Bay, and then the Chalk continues again from Punfield Cove at the northern end of Swanage Bay as far as the southern end of South Beach, Studland.

Handfast Point & Old Harry Rocks, the most eastern part of the UNESCO Jurassic Coast World Heritage Site.

Kimmeridge Clay cliffs, east of Kimmeridge Bay.

4. MAKING SENSE OF THE FOSSIL RECORD

MAKING SENSE OF THE FOSSIL RECORD

The East Dorset coast displays a succession of rocks ranging from the Jurassic of 157 million years ago through to the Palaeogene of between 23–66 million years ago. The fossil record found in these rocks is a part of our planet's intriguing history and helps to provide an interpretation of the types of palaeoenvironment in which the animal or plant once lived. The fossils found within the exposed rocks provide the scientific evidence required to help us to make some sense of the history of life, which in turn, is only a very small part the Earth's even longer and complex history; a history that began when a molten planet came into being around 4.54 billion years ago.

Whilst the rocks of Dorset are richly fossiliferous, yielding as great a range of fossils as almost any Mesozoic formation in Europe, many of these remains provide little in the way of stratigraphical information. For this, we rely on the ammonites, which evolved and died in sufficiently large numbers to allow an accurate regional correlation and help to form a detailed zonation.

Ammonites are relatively common fossils in the sedimentary rocks found on the Dorset coast. They evolved very rapidly and consequently each species of ammonite had a relatively short life span. Due to their worldwide geographical distribution, ammonites make excellent guide fossils for precise stratigraphy and accurate dating of rock layers. The rapidity of their evolution is the single, most important reason for their superiority over other fossils for the purposes of correlation and such correlation can be on a worldwide scale. Because of the ammonites' importance to stratigraphy, rock layers or groups of layers are often named after a particular species that is abundant within them. This is illustrated on Page 53, where ammonite zones from the Upper Kimmeridgian are shown.

Ammonites became extinct at the end of the Cretaceous Period, at roughly the same time as the dinosaurs disappeared. Their extinction has been attributed to the Cretaceous–Paleogene (K–PG) extinction event, caused by the impact of a massive comet or asteroid, in which it is estimated that 75% or more of all species on Earth vanished.

Kimmeridgian (Upper Kimmeridge Clay)	*Virgatopavlovia* **fittoni**
	Pavlovia **rotunda**
	Pavlovia **pallasiodes**
	Pavlovia **pectinatus**
	Pavlovia **hudlestoni**
	Pavlovia **wheatleyensis**
	Pavlovia **scitulus**
	Pavlovia **elegans**

The Upper Jurassic (Bolonian Stage) Upper Kimmeridgian succession in East Dorset and zonal ammonites.

STAGES	STANDARD ZONES	SUCCESSION		
PORTLANDIAN*	*Titanites* Anguiformis	PORTLAND GROUP	basal Purbeck Group	c. 10 m
	Galbanites Kerberus		Portland Stone Fm.	
	Galbanites Okusensis			38 m
	Glaucolithites Glaucolithus		Portland Sand Fm.	
	Progalbanites Albani			38 m
BOLONIAN*	*Virgatopavlovia* Fittoni		Upper Kimmeridge Clay	
	Pavlovia Rotunda			
	Pavlovia Pallasioides			
	Pectinatites Pectinatus			
	Pectinatites Hudlestoni			
	Pectinatites Wheatleyensis			
	Pectinatites Scitulus			
	Pectinatites Elegans			250 m
KIMMERIDGIAN	*Aulacostephanus* Autissiodorensis		Lower Kimmeridge Clay	
	Aulacostephanus Eudoxus			
	Aulacostephanoides Mutabilis			
	Rasenia Cymodoce			
	Pictonia Baylei			234 m
OXFORDIAN	*Ringsteadia* Pseudocordata	CORALLIAN GROUP	Sandsfoot Fm	30m
	Perisphinctes Cautisnigrae		Clavellata Fm	25m
	Perisphinctes Pumilus		Osmington Oolite Fm	31m
	Perisphinctes Plicatilis		Redcliff Fm	32m
	Cardioceras Cordatum	Ox Cl Fm	Weymouth Member	
	Quenstedtoceras Mariae			70m

The Upper Jurassic Succession in East Dorset, with Stages and Standard Zones.
Use with kind permission of the Geologists' Association.

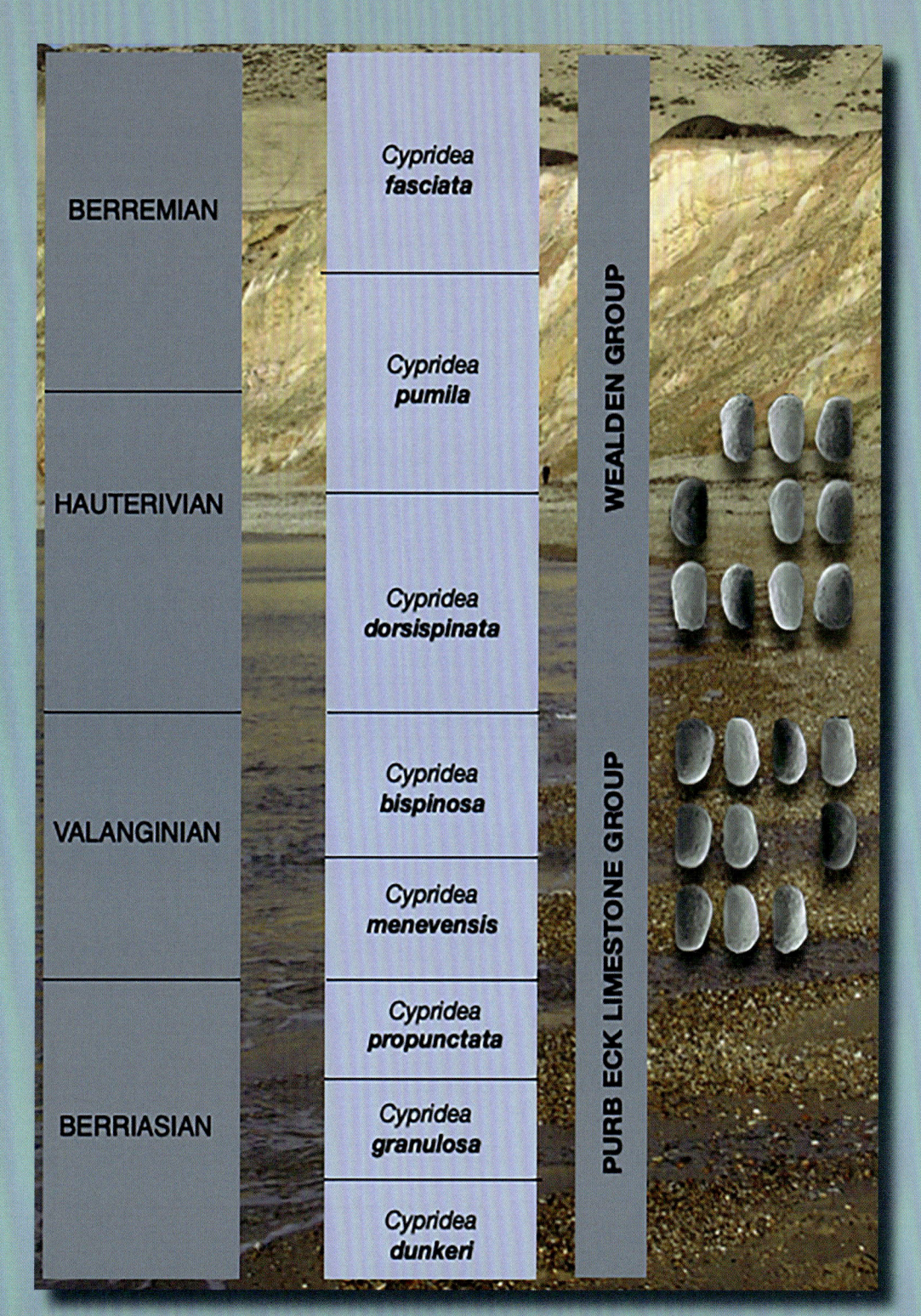

BERREMIAN	*Cypridea fasciata*	WEALDEN GROUP	
HAUTERIVIAN	*Cypridea pumila*		
	Cypridea dorsispinata		
VALANGINIAN	*Cypridea bispinosa*	PURBECK LIMESTONE GROUP	
	Cypridea menevensis		
BERRIASIAN	*Cypridea propunctata*		
	Cypridea granulosa		
	Cypridea dunkeri		

The Lower Cretaceous succession (part) in Dorset and zonal ostracods. The absence of ammonites in strata of the Wealden Group and Purbeck Limestone has required the use of ostracod fossils for zonal identification.

STAGE	ZONES	SUCCESSION
ALBIAN	*Stoliczkaia* Dispar	Upper Greensand Formation 65 m
	Mortoniceras Inflatum	
	Euhoplites Lautus	
	Euhoplites Loricatus	Gault Formation
	Hoplites Dentatus	35 m
	Douvilleiceras Mammillatum	
	Leymeriella Tardefurcata	
APTIAN	*Hypacanthoplites* Jacobi	Lower Greensand Fm — Ferruginous Sands Member
	Acanthoplites Nolani	
	Parahoplites Nutfieldensis	
	Epicheloniceras Martinioides	
	Tropaeum Bowerbanki	
	Deshayesites Deshayesi	45 m
	Deshayesites Forbesi	Atherfield Clay Mbr. 14m
	Prodeshayesites Fissicostatus	
BARREMIAN	*Cypridea fasciata*	Wealden Group
	Cypridea pumila	
HAUTERIVIAN	*Cypridea dorsispinata*	
VALANGINIAN	*Cypridea bispinosa*	
	Cypridea menevensis	715 m
BERRIASIAN	*Cypridea propunctata*	Purbeck Limestone Group
	Cypridea granulosa	
	Cypridea dunkeri	100 m

The Lower Cretaceous Succession in East Dorset, with Stages and Standard Zones.
Use with kind permission of the Geologists' Association.

STAGES	ZONES	FORMATIONS	
MAASTRICHTIAN	(absent)		
CAMPANIAN	*Belemnitella* **mucronata**	Studland Chalk Fm.	30 m
		Portsdown Chalk Fm.	60 m
	Gonioteuthis **quadrata**	Culver Chalk Fm	100 m
	Offaster **pilula**	Newhaven Chalk Formation	
	Uintacrinus **anglicus**		
SANTONIAN	*Marsupites* **testudinarius**		
	Uintacrinus **socialis**		80 m
CONIACIAN	*Micraster* **coranguinum**	Seaford Chalk Fm.	70 m
	Micraster **cortestudinarium**	Lewes Nodular Chalk Formation	
TURONIAN	*Sternotaxis* **plana**		80 m
	Terebratulina **lata**	New Pit Chalk Fm.	80 m
	Mytiloides spp.	Holywell Nodular Chalk Formation	
	Neocardioceras **Juddii**		
	Metoicoceras **Geslinianum**		50 m
CENOMANIAN	*Calycoceras* **Guerangeri**	Zig Zag Chalk Formation	
	Acanthoceras **Jukesbrownei**		
	Acanthoceras **Rhotomagense**		40 m
	Cunningtoniceras **Inerme**	Basement Bed	
	Mantelliceras **Dixoni**		
	Mantelliceras **Mantelli**		1 m

The Upper Cretaceous Succession in East Dorset, with Stages and Standard Zones.
Use with kind permission of the Geologists' Association.

The diagram on page 53 demonstrates that different types of ammonite fossils are associated with different layers of rocks in a sequence of deposition, in this case in the Upper Kimmeridgian. The stratigraphic column can therefore be divided into zones (biozones), that are characterised by one or more particular type of fossil. The sequence of these biozones in the correct order, creates a bio-stratigraphical column, which can be applied right around the world.

The use of ostracods as zone fossils

Of course, many rocks formed in environments that were not suitable for ammonites, so other creatures have to be used as zone fossils, such as the tiny ostracod. Ostracods are small, shelled crustaceans that are still living today. Their fossil record stretches right back into the Cambrian period. These little animals range in size from below a millimetre to a few centimetres but most are between 0.1 mm and 2 mm. They consist of a shrimp-like organism, encased in a bean shaped, bivalved shell.

Like other crustaceans, the hard exoskeleton surrounding the soft body of the creature must be shed periodically for it to grow. Because of this, many ostracod shells or 'valves' found as fossils are from moults, rather than because the animal has died. The animal within the shell does not fossilise.

Ostracods lived (and still do) in various aquatic environments, including fresh and brackish water. In the oceans, they inhabit both the sea floor and the planktonic zones. However, they are most commonly found on the sea floor zone and are therefore generally benthic in nature.

During the formation of the Purbeck Limestone and Wealden strata, ostracods were plentiful. These rocks formed in environments that were not suitable for ammonites but in which ostracods were abundant and so ostracods have been used as zone fossils in determining the age and type of environments that occurred during sedimentation of these rocks. This is shown in the chart on page 55.

Ostracods are by far the most abundant fossil arthropods. They are collected for many purposes and applications, including palaeoenvironmental and palaeoecological analysis, as well as dating and correlation of rock sequences, and for taxonomic and evolutionary studies. By virtue of their small size and calcified bivalve carapaces which are readily preserved, ostracods have an excellent fossil record and their valves can be recovered in large quantities from samples of sediments and sedimentary rocks.

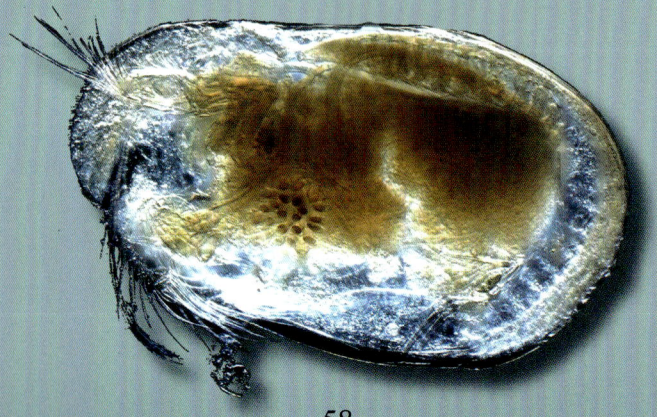

A BRIEF HISTORY OF ZONE FOSSILS

William Smith (1769–1839) is credited with creating the first detailed, nationwide geological map of any country. Smith observed the strata of the rocks whilst at work at one of Somerset's coalmines. He realised that they were arranged in a predictable pattern and that the various strata could always be found in the same relative positions. Additionally, each particular stratum could be identified by the fossils it contained and the same succession of fossil groups from older to younger rocks could be found in many other parts of England.

Smith amassed a large collection of fossils of the strata he had examined for himself, whilst working as a surveyor for the Somerset Coal Canal. This gave him a testable hypothesis, which he termed *The Principal of Faunal Succession*. And so began his search to determine if the relationships between the strata and their characteristics were consistent throughout the country.

He published his findings with many pictures from his fossil collection, enabling others to investigate their distribution and test his theories. His collection is especially good on Jurassic fossils he collected from the Cornbrash, Kimmeridge Clay, Oxford Clay, Oolitic limestone and other horizons in the sequence and resulted in another book, *Strata Identified by Organised Fossils,* in which he recognised that strata contained distinct fossil assemblages which could be used to match rocks across regions. Ammonites were particularly abundant and the different species in each rock layer were easy to identify. Different species were alive at different times, so the fossils were like time capsules, telling Smith how old each of the layers were. By noticing that fossils changed with time, he paved the way for later theories of evolution. Although unfinished, it was a fundamental work in establishing the science of stratigraphy.

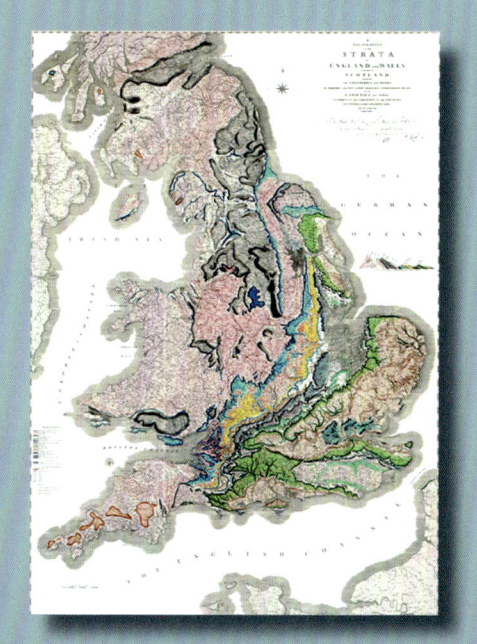

In 1801, he drew a rough sketch of what would become the first geological map of most of Great Britain. The resultant map was published in 1815 and is one of the first stratigraphical analyses to utilise paleontological indices.

William Smith laid the foundation for stratigraphy in England. Later, his pioneering work was to be continued by others. Of note were Carl Albert Oppel (1831–1865) and Alcide d'Orbigny (1802–1857).

Carl Albert Oppel provided a detailed zonation of the Jurassic by the use of ammonites and was eventually able to subdivide the Jurassic into 33 different zones. Oppel had devoted his life to the study of fossils and the examination of the strata of the Jurassic period deposits. He is now considered to have founded the study of zone stratigraphy and the use of index fossils, a term he created, to compare the different strata.

He also established the Tithonian stage, for strata (mainly equivalent to the Portland and Purbeck Beds of England) that occur on the borders of the Jurassic and Cretaceous.

The extraordinary merit of Oppel's work has not only been the demonstration that fossils can be used to sub-divide sedimentary sequences into zones, but these in turn, might be organised in higher chronostratigraphical units. The zone for Oppel is characterised by the distinctive fossil content and his view strongly influenced the development of the standard chronostratigraphical scale for about one century.

Ammonite zones therefore identify periods of time lasting approximately 250,000 years. Increased granularity has been provided by the use of ammonite Sub-zones, which can define considerably less time and certainly offer much finer resolution when trying to figure out the history of time as it is recorded in the rock layers.

To the fossil enthusiast or collector, the relationship between the ammonite fossil and the rock from which it was found can offer a reasonably accurate means of dating the specimen and in helping to determine the palaeoecosystem in which the creature once lived.

In 1849, French naturalist Alcide d'Orbigny published a closely related *Prodrome de Paléontologie Stratigraphique*, intended as a *Preface to Stratigraphic Palaeontology*, in which he described almost 18,000 species, and with biostratigraphical comparisons erected geological stages, the definitions of which rest on their stratotypes.

He also described the geological timescales and defined numerous geological strata, still used today as chronostratigraphic reference such as Toarcian, Callovian, Oxfordian, Kimmeridgian, Aptian, Albian and Cenomanian.

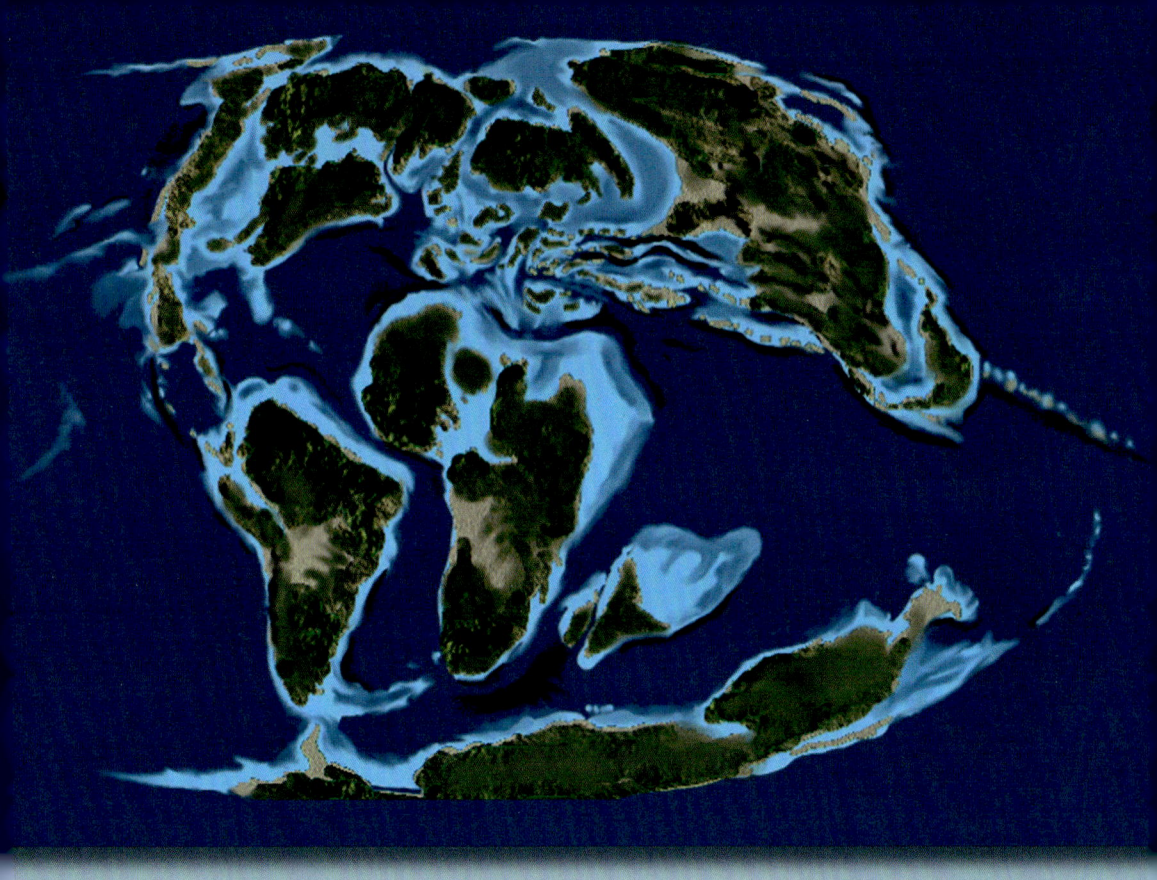

LATE JURASSIC SERIES

The late Jurassic in Britain brought forth a period of further change to the position of the continents, as the landmasses drifted to positions more familiar to us today. Sea levels continued to rise and the seas saw the further rise in the size and diversity of life forms, all of which thrived and evolved. Marine fauna increased in abundance as their environment expanded and deepened. The period of globally high sea levels are demonstrated in the predominant lithology of the Kimmeridgian mudrocks; the most extensive of Mesozoic deposits.

During the Middle Jurassic, the region was deeply buried beneath marine sediments but the Late Jurassic saw a brief return to land, upon which plants and dinosaurs flourished under the more than favourable climatic conditions. The Late Jurassic rocks of East Dorset bear testimony to these changes and in which the fossil record demonstrates the life on Earth at this time.

THE UPPER JURASSIC SERIES IN EAST DORSET

THE KIMMERIDGE CLAY FORMATION

The Kimmeridge Clay Formation exposed along the East Dorset coast is a typical sedimentary mudstone deposit, formed in shallow marine conditions around 152–157 million years ago. The rock is detrital, ranging from coarse to fine-grained, forming interbedded sequences. In East Dorset, the Lower and Upper Kimmeridge Clay extends along the coast from Gad Cliff at Brandy Bay, eastwards to St. Aldhelm's Head.

TERRESTRIAL REPTILES OF THE KIMMERIDGE CLAY FORMATION

Ischyrosaurus

Dacentrurus

Duriatitan

Pelorosaurus

Cuspicephalus

Cetiosaurus

Rhamphorhynchus

Juratyrant

Germanodactylus

MARINE REPTILES OF THE KIMMERIDGE CLAY FORMATION

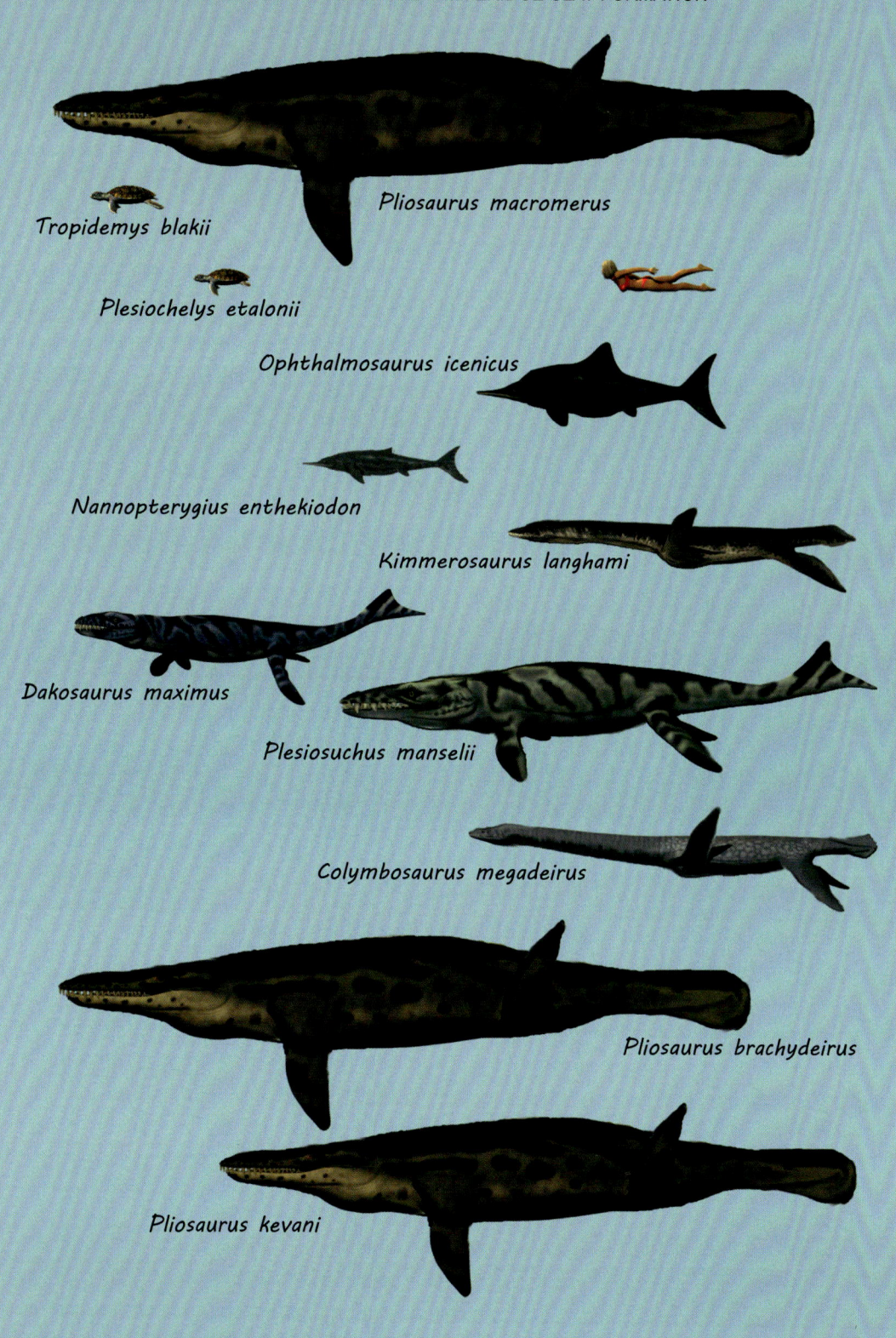

Pliosaurus macromerus

Tropidemys blakii

Plesiochelys etalonii

Ophthalmosaurus icenicus

Nannopterygius enthekiodon

Kimmerosaurus langhami

Dakosaurus maximus

Plesiosuchus manselii

Colymbosaurus megadeirus

Pliosaurus brachydeirus

Pliosaurus kevani

Cliffs east of Kimmeridge Bay, looking west.

Unlike the complete thickness of Lower Kimmeridge Clay exposed in the southern section of the coast at Ringstead Bay, in the east only the upper part of the Lower Kimmeridge Clay Formation appears, followed by the full type section of Upper Kimmeridge Clay around Kimmeridge. The exposures reveal a formation from which the fossil fauna includes an abundant and diverse group of vertebrates, which swam in these Late Jurassic seas; crocodiles, turtles, pliosaurs, ichthyosaurs, plesiosaurs, cartilaginous and bony fishes, a great number of invertebrates, including ammonites, belemnites and crinoids.

The Kimmeridgian Stage marks a period of widespread marine mudrock deposition in northwest Europe and the thick sequence of clays and bituminous shales is considered to have been deposited in calm bottom waters. Most of the sediments are terrestrially derived, indicating considerable erosion from a nearby landmass. Fossilised wood is common in the formation. It is no surprise that several species of dinosaurs and fragmentary pterosaurs occur in the formation.

The formation's maximum thickness of around 140 m is reached at the type section in the Purbeck and Kimmeridge Bay areas and comprises a series of blocky clays and fissile shales, with intermittent oil shales, cementstone bands, and layers rich in septarian nodules. The Dorset type section of the Kimmeridge Clay Formation can be divided into 13 ammonite biozones, from the *Pictonia baylei* Zone at the base to the *Virgatopavlovia fittoni* Zone at the top.

THE PORTLAND GROUP

There are major differences between the Portland Group succession on the Isle of Portland and that on the Isle of Purbeck in East Dorset. Whereas on Portland, the Portland Sand Formation is succeeded by the Portland Clay at the base of the Portland Stone, followed by the Basal Shell Bed, on the Isle of Purbeck, the Portland Clay and Basal Shell Bed are not easily accessible.

Of course, **the Portland Stone Formation** has been worked in quarries along the East Dorset coast for centuries and the succession of the Portland Group is exposed in its entirety in eastern Purbeck. The sequence is clearly seen at Seacombe. At St. Aldheim's Head, in the old quarry face, the top of the Portland Stone is visible. The Portland Stone Formation is divided into two members: the **Portland Cherty Member** and the **Portland Freestone Member** above. The Cherty Member is mostly comprised of limestones, with nodules of chert. Here, ammonites such as *Titanites giganteus* occur, some reaching large diameters.

Portland Stone cliffs at St. Aldhelm's Head, Isle of Purbeck.

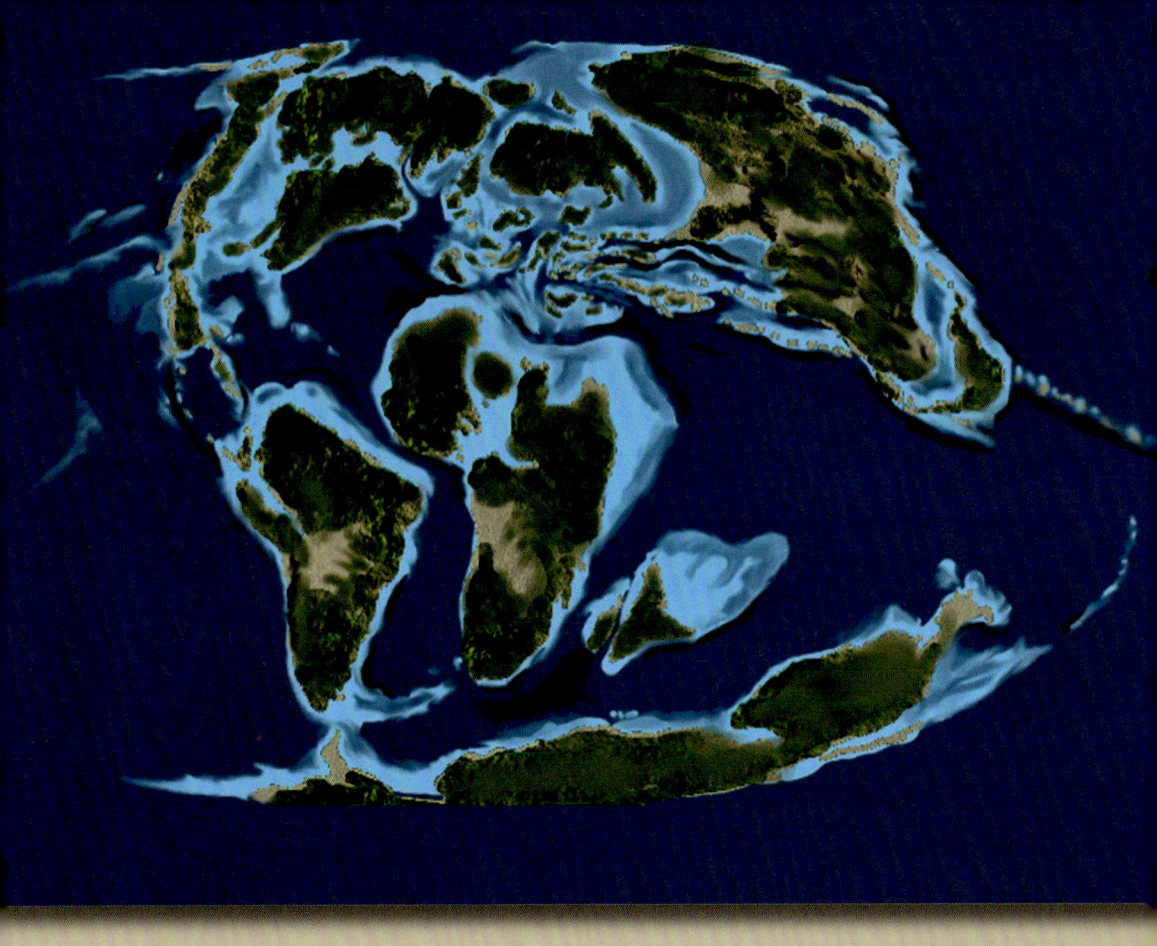

CRETACEOUS SERIES

The Cretaceous is a geological period that lasted from about 145–66 million years ago (mya). It is the third and final period of the Mesozoic Era, as well as the longest.

The Cretaceous rocks found along the East Dorset coast are varied: limestones and shales of the Purbeck Group, sands, clays and grits of the Wealden Beds, sands from the Lower Greensand, Gault Clay and the ubiquitous Chalk.

It is generally accepted that for most of Early Cretaceous time the areas of southern Britain were mainly emergent, following a significant sea-level fall that commenced very late in the Jurassic. The modern continents having formed, the Cretaceous saw the formation of the Atlantic Ocean, gradually separating northern Scotland from North America. The land underwent a series of uplifts to form a fertile plain. After 20 million years or so, the seas started to flood the land again until much of Great Britain was again below the sea, though sea levels frequently changed. The Upper Cretaceous saw the deposition of the Chalk over much of the British Isles region, with progressively inundated land areas.

LOWER CRETACEOUS SERIES IN EAST DORSET

THE PURBECK LIMESTONE GROUP

The Purbeck Limestone is a succession of limestones and marls, forming a sedimentary bedrock of approximately 135–152 Ma. The sediments are fluvial, palustrine and shallow marine in origin. They are detrital, forming deposits reflecting the channels, floodplains and deltas of a river in a coastal setting, with periodic inundation from the sea.

It comprises two formations: the **Lulworth Formation**, the first few metres of which is Jurassic, accumulated in a closed lagoon, whose salinity varied (sometimes becoming freshwater) and the rest of the Formation along with the **Durlston Formation** above is of Cretaceous age. The latter Formation was formed in an open system, which allowed marine species to colonise the lagoon, depending on the suitability of salinity.

A scene from early Wealden times, with an iguanodontid dinosaur and small mammals, which were evolving and whose small teeth and bones are found in both the Purbeck limestone and Wealden rocks.

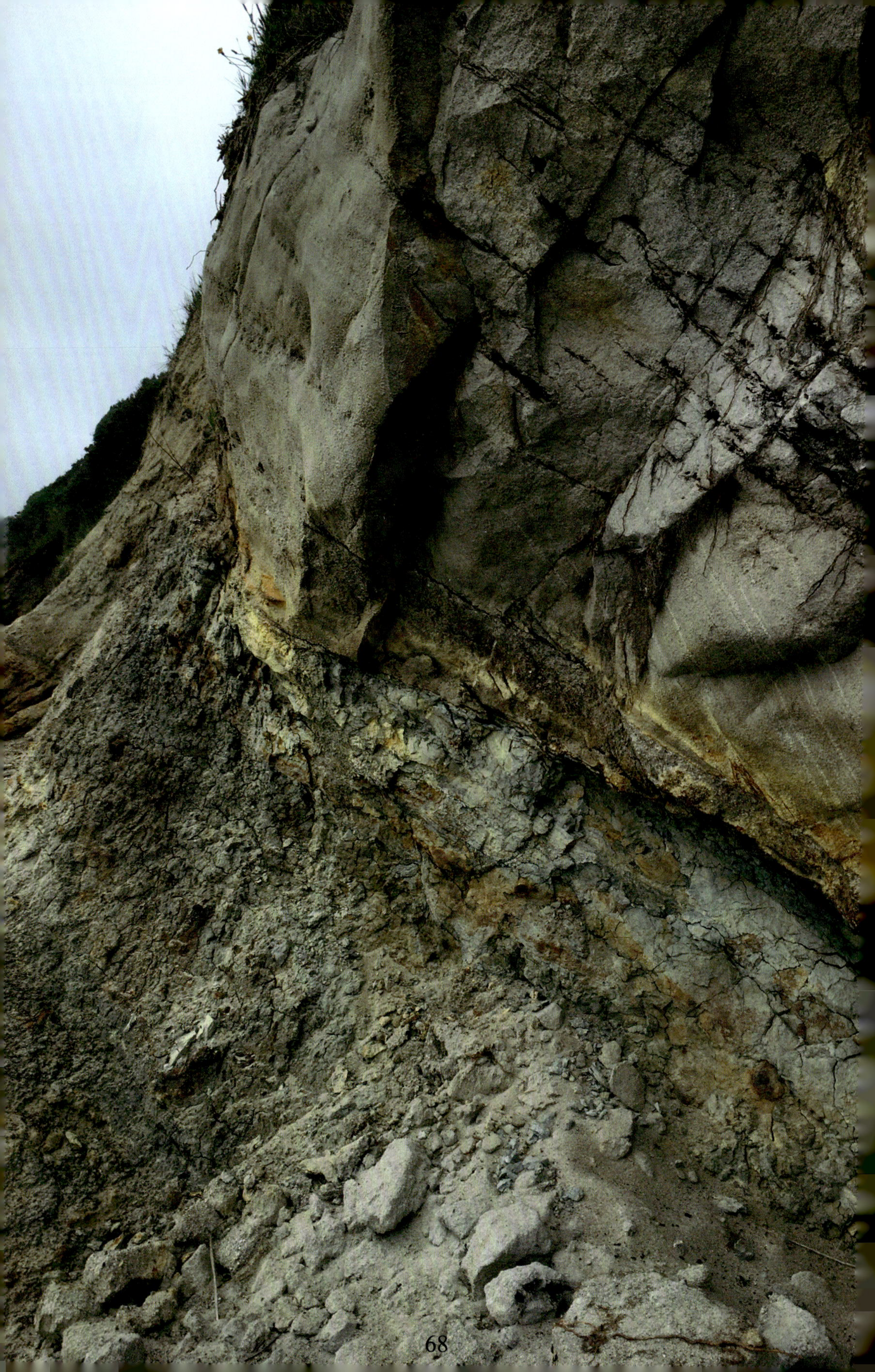

The marls and sandstones of the Wealden Group, seen in the colourful cliffs at Worbarrow Bay. Fossils are generally not common in the Wealden rocks of Dorset (unlike the Wealden clays and sandstones on the Isle of Wight), however, the soft Wealden mudstones, sandstones and grits at Swanage Bay do contain *Unio* shells and the occasional remains of dinosaurs, usually an isolated bone. There is, however, evidence of plant material and wood as shown in the photos on the right, taken at the northern part of Swanage Bay.
Photos by Viv Field.

Opposite page: Wealden clays at Swanage Bay. The formation is composed of alternating sands and clays. The sands were deposited in the flood plains of braided rivers and clays mostly in lagoonal conditions that varied from freshwater to saline. Photo by Viv Field.

THE WEALDEN GROUP

The 'regime of swamps and still muddy lagoons' came to an end and caused by Earth movements accelerated subsidence, it resulted in a rapid change to sedimentation in the Purbeck lagoon, bringing coarse sediment into the region. The formation of an extensive, shallow basin across southern England saw rivers flowing into it from all directions, bringing vast amounts of detritus and floating vegetation. These were spread out along the bottom of the basin as false-bedded sands, marls and clays, with bands of coarse grits. As a result, the Wealden Group consists of sands and clays, with ironstones at some horizons and the marls and sandstones are well represented in the red, orange, yellow and purple cliffs at Worbarrow Bay. Disappointingly, the Wealden strata in Dorset is mostly unfossiliferous, unlike the Isle of Wight.

THE LOWER GREENSAND FORMATION

The Lower Greensand Formation ends a period of non-marine deposition and begins a long cycle of marine transgression and sedimentation, although in Dorset the succession is still not fully marine. It is not significantly fossiliferous in Dorset, although the section at Swanage, in the Punfield Marine Bed, can yield ammonites (*Deshayesites punfieldensis* and *Parahoplites* sp.), bivalves and gastropods. In East Dorset, the Lower Greensand begins with a basal pebble bed containing phosphatic nodules and fish teeth, as seen at Swanage. However, fossils are difficult to collect, owing to fragile preservation and slipping of the clays at the only exposure at Punfield Cove, at the north corner of Swanage Bay. The overlying Ferruginous Sands are unfossiliferous.

The beach at the northern end of Swanage Bay, referred to as Punfield Cove.

Micraster sp. echinoid in Upper Greensand Formation at St. Oswald's Bay.

THE GAULT & UPPER GREENSAND FORMATIONS

The best rocks of the Middle Albian sequence are found at Worbarrow Bay, where the Gault Formation attains a thickness of some 34.5 m and where a succession of dark green or black clays occur. Fossils are not common and unlike exposures elsewhere in Britain, consist mainly of worm tubes and bivalves.

Across the Isle of Purbeck, the Upper Greensand Formation consists of green, glauconitic sands. At present, the Upper Greensand suffers from good exposures and considerable parts of the outcrop are often hidden under slipped masses of Chalk, although some parts are exposed at the base of Ballard Cliff at Swanage Bay. The rocks were formed under shallow water, marine conditions and form inter-bedded sequences of fine grained to coarse detrital material, deposited 94–113 mya.

Echinoid in Upper Greensand at Swanage. Photo by Viv Field.

Lower Cretaceous scene with a *Carcharias taurus*, a sand tiger shark with a *Leptocleidus* pliosaur, a *Lepidotes* shoal and heteromorph ammonites. Artwork by Andreas Kurpisz.

THE UPPER CRETACEOUS SERIES IN EAST DORSET

THE CHALK GROUP

The Chalk Group has undergone a considerable change to the subdivisions that previously existed – namely the Lower, Middle and Upper Chalk, still referred to by many. The Chalk Group is now divided into two Subgroups and nine formations in Dorset. The lower is the **Grey Chalk Subgroup,** which essentially replaces the Lower Chalk and begins at the base of the West Melbury Marly Chalk Formation and is succeeded by the Zig Zag Chalk Formation.

The White Chalk Subgroup starts with the Plenus Marls Member of the Holywell Nodular Chalk Formation and is followed by the New Pit Chalk Formation, the Lewes Nodular Chalk, the Seaford Chalk Formation, Newhaven Chalk Formation, Culver Chalk Formation, the Portsdown Chalk Formation and finally the Studland Chalk Formation.

The cliff at Ballard Down forms the northern edge of Swanage Bay and is nearly 100 metres of Grey and White Chalk Subgroups.

TERTIARY (PALAEOCENE) SERIES

The Palaeocene rocks of Dorset are restricted to the far east of the county. Traditionally, the lowest formation has been known as the **Reading Formation** and is still favoured by many, despite a re-mapping of the area by the Geological Survey. They rest with a disconformity on the White Chalk Subgroup.

The various clays and sands of the **Harwich Formation** and **London Clay Formation** lie above this, succeeded by the Bracklesham Group (composed of a lower **Poole Formation** and an upper **Branksome Sand Formation**. None are particularly fossiliferous and are generally poorly exposed. The Poole Formation is exposed at Studland Bay, as a basal member known as Redend Sandstone and as the Creekmoor Clay above it, which yields abundant fossil plant material.

The succeeding Barton Group comprises four formations: the **Boscombe Sands Formation,** the **Barton Clay Formation**, the **Chama Sand Formation** and the **Becton Sand Formation,** of which only the latter two crop out in Dorset.

Lower Cretaceous marine life with ammonites *Aconeceras* and *Douvilleiceras*, with fish *Gyrodus*, *Dapedium*, *Eomesodus* and *Macromesodon*. Artwork by Andreas Kurpisz.

5. FOSSIL COLLECTING EXCURSIONS IN EAST DORSET

Mupe Rocks, looking east towards Mupe Bay.

EXCURSIONS NEAR DURDLE DOOR

Bat's Head to St. Oswald's Bay, Dungy Head & Lulworth Cove

CHALK GROUP
PURBECK GROUP
UPPER GREENSAND
GAULT

St. Oswald's Bay, looking west towards Durdle Door.

BAT'S HEAD & SWYRE HEAD

The section of coast east of White Nothe, the Chalk promontory at Ringstead Bay is only accessible by means of a small boat and the range of fossils to be collected remain beyond the means of most.

The beach to the east of Bat's Head and west of Durdle Door, beneath the vertical cliff of Swyre Head is accessed from west of Durdle Door. The photo (below) shows the headland of Bat's Head in the distance, which displays a vertical section through the White Chalk Subgroup, with New Pit, Lewes Nodular and Seaford Chalks displayed in the cliffs nearest to the camera.

Bat's Head is almost entirely in the *Micraster cortestudinarium* Zone, with the tip in the *Sternotaxis planus* Zone. These zones are now classified within the Lewes Nodular Chalk Formation of the White Chalk Subgroup. The Chalk echinoids can sometimes be found on the beach, having eroded out of the rock or found as flint casts. The Chalk along this section of beach is uncharacteristically hard and fossils are not particularly forthcoming. The beaches on both sides of Durdle Door are popular tourist beaches and collecting might be best suited to the times of year when less frequented and when erosion rates are highest.

The beach below the White Chalk cliffs of Bat's Head, Swyre Head with Scratchy Bottom between them. Fossils are *Sternotaxis planus* (left), *Micraster cortestudinarium* (centre) and *Inoceramus*, a bivalve commonly found in the Chalk.

In the section, fossiliferous beds of Upper Greensand Formation occur, overlain by the Chalk Group. The base rests on the Upper Greensand; the Cenomanian Basement Bed – a glauconitic limestone, with phosphatised pellets, which weather proud of the rock. These should be examined for rolled ammonites and other fossils. The echinoid, *Holaster* and ammonites (including *Schloenbachia*, *Calycoceras* and *Mantelliceras*) and heteromorph ammonoids, *Scaphites* and *Turrilites* occur in the top bed. Above is the Grey Chalk Subgroup, followed by the White Chalk Subgroup, with the New Pit, Lewes Nodular and Seaford Chalk Formations visible towards Scratchy Bottom. Echinoid fossils are common, so expect *Micraster coranguinum*, *Echinocorys* and *Conulus*.

ST. OSWALD'S BAY, DUNGY HEAD & DURDLE DOOR

St. Oswald's Bay lies just to the east of Durdle Door. This is an impressive section of coast, with good exposures from the Jurassic Portland Sand, through the Portland Stone, Purbeck Formation, Wealden, Gault and Upper Greensand through to the Chalk. To get to St. Oswald's Bay, walk up the short hilly road westward of Lulworth Cove, past Stair Hole and the large red-brick houses. At the end of the tarmac road there is an easy footpath ahead leading down to the beach at St. Oswald's Bay, near Dungy Head. A short climb down some muddy steps and an old landslide will get you to the beach.

Upper Greensand Formation at St. Oswald's Bay, with *Ostlingoceras*, a coiled heteromorph. Photo by Viv Field.

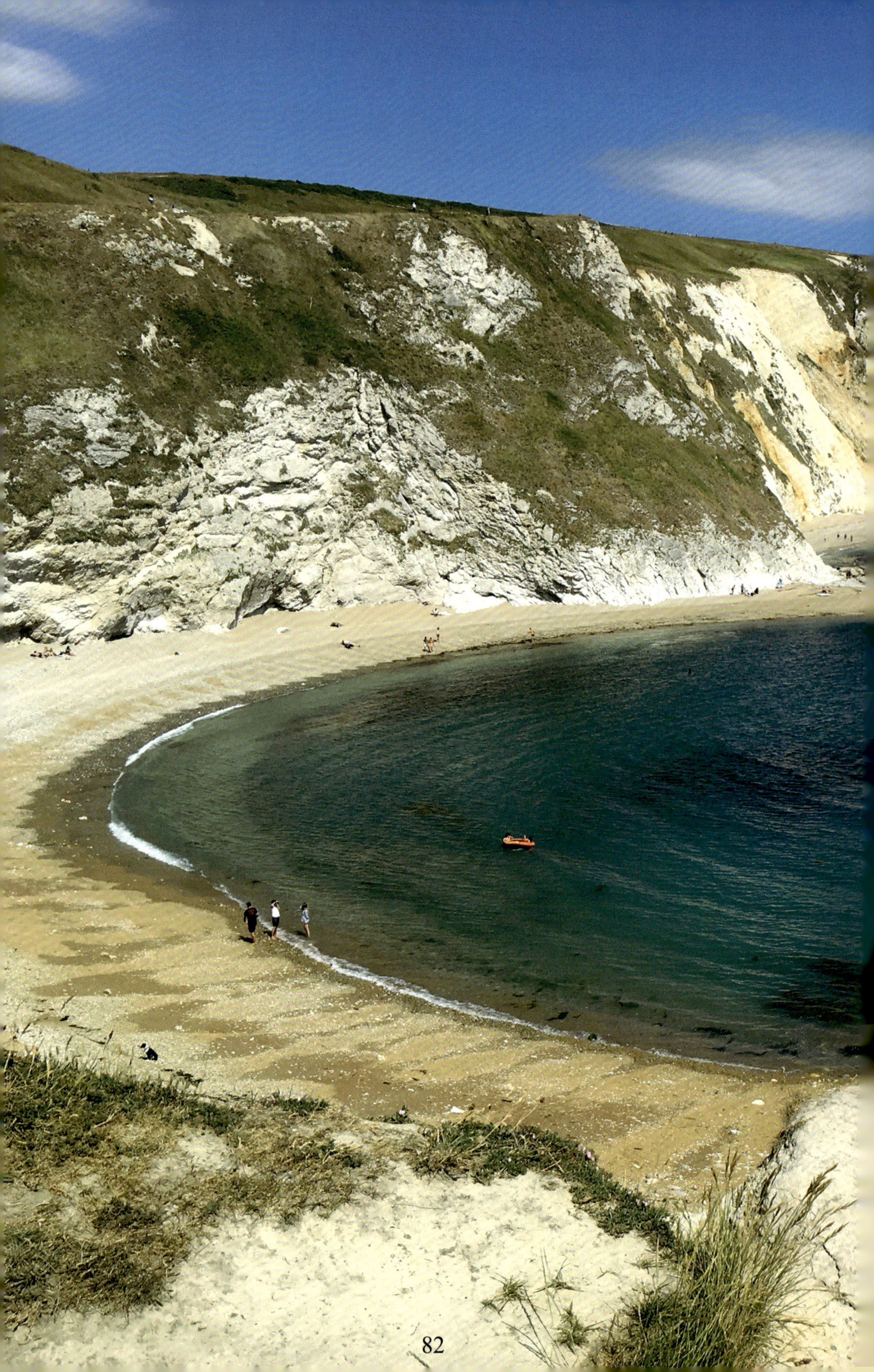

Man O' War Cove and St. Oswald's Bay looking east.

The faults and other structures in St. Oswald's Bay are particularly interesting. There are complications of strike faulting in steeply dipping and overturned strata. For the most part, although of significant interest to the structural geologist, the section is not highly fossiliferous, save for a section close to the path down to beach level. This is comprised of Upper Greensand Formation, consisting of dark green glauconitic silts with harder beds in places. The rock is fossiliferous and from which bivalves, brachiopods, worm casts and echinoids may be collected. Ammonites (including the coiled heteromorph, *Ostlingoceras*) also occur here, along with shark teeth, (including the uncommon *Ptychodus*) and insects.

The basal Gault Clay, with its pebble bed, is relatively scarce and the site may need to be revisited when exposures are more favourable but the rounded bed of quartz and phosphatised pebbles also contain phosphatised moulds of bivalves and occasional reptile teeth.

The Upper Greensand Formation at St. Oswald's Bay yields abundant specimens of the bivalve *Aequipecten aspera*.
Found and photographed by Viv Field.

Tooth of *Ptychodus* sp., a hybodont shark.

Tooth of *Scapanorhynchus* sp., a deep-water species of shark.
Found and photographed by Viv Field.

MAN O' WAR COVE & DURDLE DOOR

Man o' War Cove is the western most part of St. Oswald's Bay and access is best by continuing to walk west from the previous location. Again, quite fossiliferous beds of Upper Greensand Formation occur, overlain by the Chalk Group. The base rests on the Upper Greensand; the Cenomanian Basement Bed – a glauconitic limestone, with phosphatised pellets, which weather proud of the rock. These should be examined for rolled ammonites and other fossils. The echinoid, *Holaster* and ammonites (including *Schloenbachia*, *Calycoceras* and *Mantelliceras*) and heteromorph ammonoids, *Scaphites* and *Turrilites* occur in the top bed.

Above is the Grey Chalk Subgroup, followed by the White Chalk Subgroup, with the New Pit, Lewes Nodular and Seaford Chalk Formations visible towards Scratchy Bottom. Echinoid fossils are common, so once again expect to find *Micraster coranguinum*, *Echinocorys* and *Conulus*.

At Man o' War Cove take in the view of the famous arch of Durdle Door, comprised of Portland Stone on its seaward side and basal Purbeck Group on the landward side. The top of the arch has evidence of circular tuffaceous impressions that once enclosed tree stumps and not dissimilar to those seen at the Fossil Forest at Lulworth.

Man o' War Cove showing Cenomanian ammonites, *Calycoceras* (left) and *Schloenbachia* (right).

The large Portland Stone islet in the bay is the Man o' War. The cliffs expose the rocks of the Purbeck Group, with the cinder Bed displayed at the foot of the steps. The section has been heavily compressed, faulted and overturned. The Wealden Group sands can also be examined at this point, with plenty of plant debris and particles, much of it in the form of charcoal; evidence of wildfires in the vegetation surrounding the Wealden lake. The more silty bands within the rock contain identifiable plant remains, such as the fronds of *Bennettitales* (also known as cycadeoids), an extinct order of seed plants.

DUNGY HEAD

Returning towards Dungy Head in the east, the **Portland Stone Formation** is exposed, particularly on the western side of the promontory. Within the beds, the Basal Shell Bed may reveal ammonites. Blocks of Portland Stone litter the beach and traversing this area is difficult, with danger of cliff falls. Traversing around Dungy Head, a section of the underlying **Portland Sand Formation** has, in the past, revealed the ammonite *Virgatupavlovia*. Higher beds have revealed *Progalbanites* and *Epivirgatites* of the *albani* Zone.

Epivirgatites (left) and *Glaucolithites* (right) ammonites from the Portland Sand Formation exposed at Dungy Head.

In the **Portland Sand,** at the top, occurs Black Sandstones, which is part of the Gad Cliff Member. The ammonite *Glaucolithites* occurs here, with abundant bivalves in the Cast Bed (also part of the Gad Cliff Member) below. *Glaucolithites* occurs here also. The *Exogyra* Beds with *Nanogyra nana*, the serpulid *Glomerula gordialis* and *Plicatula boisdini*. The collector may have to suffice with specimens from fallen blocks, as this section is not best climbed!

The small bivalve *Nanogyra nana*.

The **Portland Cherty Series** is also well seen at Dungy Head and both the upper part (or Dancing Ledge Member) and the lower part (the Dungy Head Member), contain much sponge spicule (*Rhaxella*) wackstone. The conspicuous thin marker bed known as Puffin Ledge is easily recognised and is characterised by *Thallassinoides* burrows. *Titanites* ammonites occur here.

STAIR HOLE & LULWORTH COVE

Stair Hole is a small cove situated immediately to the west of Lulworth Cove. There is an outer barrier of steeply dipping Portland Stone and basal Purbeck limestones (the Caps). Stair Hole demonstrates erosional effects leading to a breaching of the Portland Stone cliffs in two places and where the Purbeck Group behind them has been extensively eroded also. The Wealden Group is seen slipping down into the formed embayment.

Stair Hole, Lulworth. The Lulworth Crumple of the Purbeck Group is due to tectonic compression during the Alpine deformation.

In the past, pieces of *Tempskya* tree fern from the base of the Wealden strata have been found at Stair Hole. On the east side of Lulworth Cove, the strata are from the Purbeck Group (Upper Purbeck), Durlston Formation, and represent a freshwater lacustrine (lake) origin. Here, the pond-snail *Viviparus*, the freshwater bivalve *Unio*, fish remains (e.g., *Pycnodus, Lepidotus*) and ostracods are common in these beds. Turtle bones also occur in Upper Purbeck glauconitic limestones, along with the bivalve *Unio*. The brown turtle bones contrast well with the grey rock and teeth, scutes and bones of crocodiles can also be found in this unit at various places.

Typical bone fragment, possibly from a humerus of a turtle, quite probably *Hylaeochelys* sp., a shell-based genus of primitive turtle known from the Lower Cretaceous Purbeck Group. Specimen found by John Aldron.

Reconstruction of the turtle *Hylaeochelys* sp. Artwork by Craig Chivers.

Fossil Forest at West Lulworth

FOSSIL FOREST

The Fossil Forest at West Lulworth lies on a ledge approximately 25 metres above the sea. It provides an insight into conditions prevailing in the Late Jurassic around 135 million years ago. The ring-shaped structures, up to two metres across, are the moulds of gymnosperms; early coniferous trees, which died after being encased in tuffaceous limestone. The trees were mostly upright, although the elongated coffin-shaped moulds are evidence of trees that had leaned and eventually fallen. The Fossil Forest, which can be reached by walking east, is only accessible when the army ranges are closed.

LULWORTH COVE

The high cliffs at the back of the Cove are in the Chalk Group. Most of the fossils at Lulworth are found from the Grey Chalk Subgroup and the White Chalk Subgroup at Bindon Hill (in the middle of the cove). However, they can also be found in the Upper Greensand Formation, which is exposed near the old limekiln.

Lulworth Cove.

Purbeck Group
The **Hard Cockle Member** here consists of lagoonal limestones with *Protocardia purbeckensis*, a euryhaline cockle.

The **Soft Cockle Member** may contain insect remains.

The oyster-bearing **Cinder Bed** in involved in small folds. The **Intermarine Member**, brackish water *Neomiodon* limestones are much thinner here than in the Swanage area where they are extensively quarried. The bivalve *Neomiodon medium* is abundant. Look for the **Corbula Member**; you should see the small bivalve *Corbula durlstonensis*. The **Chief Beef Member** is obvious from the large veins of fibrous calcite. You will clearly see the brown stained and green *Unio* **Member** with a large slab dipping north, where you may also find fish teeth (*Lepidotus*). To the north east on the beach look for *Unio* in the Upper Purbeck strata and for Purbeck marble with *Viviparus*.

Upper Greensand
The Upper Greensand is about 40 metres thick at Lulworth Cove and consists of calcite-cemented sand and sandstone with marine fossils.

The calcite-cemented sandstone of the Exogyra Rock, which is about one-metre-thick occurs in the middle of the succession and near the entrance to Lulworth Cove. This bed is conspicuous because it projects in the cliffs and is full of fossils, particularly *Exogyra obliquata*, and small echinoids (e.g. *Catopygus carinatus* (left), *Salenia petalifera* (right) and *Discoidea*) are fairly common in some exposures.

EXCURSIONS NEAR WORBARROW BAY

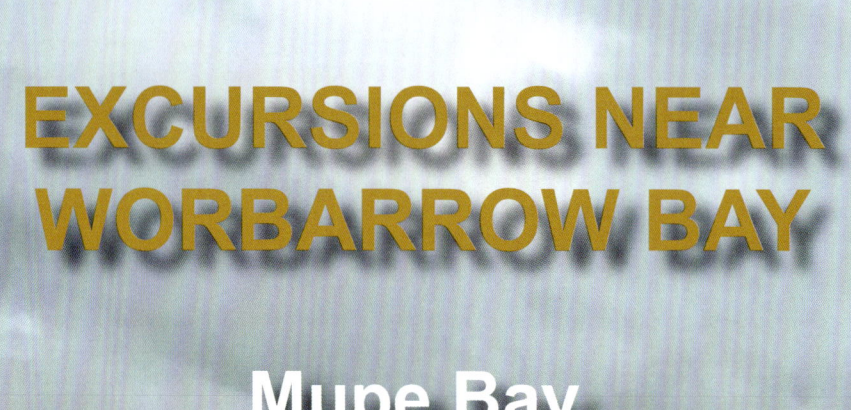

Mupe Bay, Worbarrow Bay, Brandy Bay & Gad Cliff

CHALK GROUP
PURBECK GROUP
UPPER GREENSAND
GAULT

Worbarrow Bay, looking west from the coastal path near Worbarrow Tout.

MUPE BAY

Mupe Bay lies within the Army Firing Range and access is limited to certain times. It is usually open at weekend but you will need to check the well-displayed notices in the area.

The rocks at Mupe Rocks are formed of Portland Stone Formation with the lower part of the Purbeck Limestone Group, being Durlston and Lulworth Formations. Descend the steps to Mupe Bay and turn right, to see the ledges along the foreshore. Here, at low tide, the Cinder Bed is composed of *Praeexogyra distorta* (shown above). The cliffs are of the softer Wealden Group (principally mudstones of shallow marine origin and into which the embayment is cut) at the western end of the bay, with unfossiliferous Lower Greensand. The section is badly obscured by Chalk talus and no fossils have been recorded from this location. Gault and Upper Greensand Formation are followed by the Chalk, with Zig Zag, Holywell Nodular, New Pit, Lewes Nodular and Seaford Chalk Formations displayed.

The majority of fossils to be found here will be from the Chalk and include *Micraster coranguinum* (Seaford Chalk Formation) and *Micraster cortestudinarium* (Lewes Nodular Chalk Formation) but exposures appear to be quite unfossiliferous. The collapsed cliffs are quite unstable and rock falls are commonplace.

From the west of the bay, the Cenomanian strata is unfossiliferous and fossils only start to appear in the Turonian further east. Better collecting can be had from the *planus* Zone boulders on the beach.

The small embayment at Arish Mell, is now totally inaccessible.

Mantelliceras sp. ammonite (Cenomanian Basement Bed of the Chalk Group) and *Callihoplites* sp. and *Stoliczkaia* sp. (Upper Greensand Formation) of Worbarrow Bay.

WORBARROW BAY

The geology at Worbarrow Bay is quite complex but essentially the rock types are a mix of Wealden, Lower Greensand, Gault, Upper Greensand and Chalk, with much faulting and with much of the outcrops hidden by slipped masses of Chalk.

Access to Worbarrow Bay is best gained by parking in Tyneham, the deserted village and from here following a track leading down to the beach. On Worbarrow Tout, the Purbeck Limestone Group is well exposed. In the eastern part of the bay, the sequence of strata is quickly succeeded by the Wealden Group, comprising mainly brightly coloured sands and clays.

On the whole, fossils are rare but some ammonites do occur in the Chalk and fallen blocks of Upper Greensand Formation. From the latter, these include *Anahoplites*, *Callihoplites*, *Hysteroceras*, *Durnovarites* and *Stoliczkaia*.

The coloured sands of the Wealden Group at Worbarrow Bay. The clays and sands of the Wealden Group, displayed in the extensive cliffs at Worbarrow Bay are mostly unfossiliferous.

A nautilus found in the Chalk at Worbarrow Bay. Photo and found by Hashimoto Tadanori.

Ripple marks in the Durlston Formation at Worbarrow Bay. Photo by Viv Field.

Eastward, in the cliffs the Chalk commences with a Basement Bed; a distinctive bed 1.2 metres-thick, of soft, brownish, sandy chalk. In the pockets of this many white, clayey casts of small ammonites have been found. They include (left to right): *Calycoceras newboldi*, *Calycoceras gentoni*, *Scaphites aequalis* and *Scaphites obliquus*.

Dinosaur prints in the Durlston Formation (Purbeck Group) at Worbarrow Bay. Photo by Viv Field.

Fish in the Grey Chalk Subgroup at Worbarrow Bay. Photo by Viv Field.

Gad Cliff, from the ledges beyond Broad Bench. Photo Viv Field.

BRANDY BAY & GAD CLIFF

The impressive cliff that dominates Brandy Bay is Gad Cliff, composed of overhanging Portland Stone, above the marls of Portland Sand and with Kimmeridge Clay at the base. When the firing ranges are open (see published timetable) it is possible to view the coast eastwards, to the Chalk cliffs at Mupe Bay and to the Isle of Portland, from the top of Gad Cliff.

Despite Brandy Bay being fossiliferous, access to the beach and ledges is not permissible. In some respects, although disappointing, the tide is a limiting factor and when the tide is rising, it is not possible to proceed any significant distance beyond Kimmeridge Bay. In any case, Brandy Bay is in the Army firing range and, even when the range walks are open, the bay remains closed off. Fossils are mostly of the usual ammonites and in the bay are species of *Pectinatities* from the Upper Kimmeridge Clay, but some other genera also occur, such as *Aspidoceras*, as well as various bivalves including *Myophorella clavellata*.

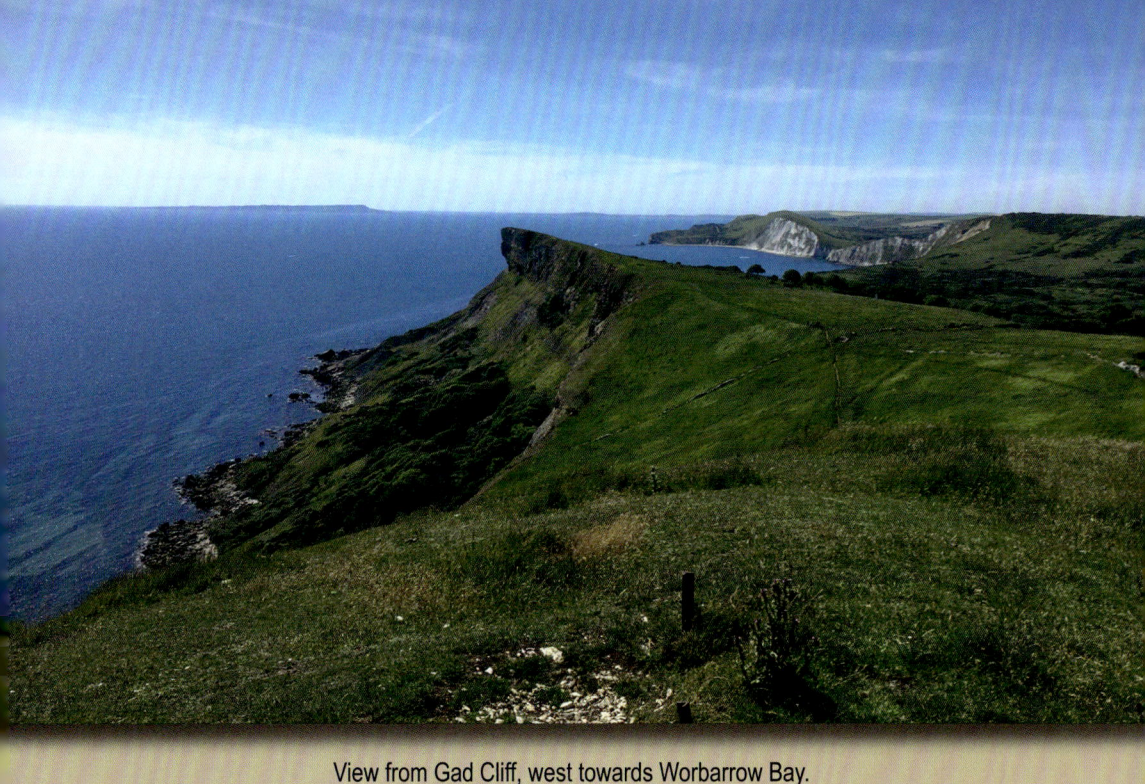

View from Gad Cliff, west towards Worbarrow Bay.

The area to the west of Kimmeridge Bay is best accessed from Kimmeridge Bay itself. Access to Hobarrow Bay and Brandy Bay is restricted, due to being within the Army firing range and the shore is not for public access, even when firing is not taking place. Nonetheless, it is possible to walk around Kimmeridge Bay to Broad Bench, a ledge formed by the Flats Dolomite Bed.

Fossils here are much the same as found at Kimmeridge Bay, with partly or wholly crushed ammonites of the genus *Aulacostephanus* and Lower Kimmeridge Clay bivalves, such as *Protocardia* and *Nanogyra virgula* often found in broken slabs of Kimmeridge shale.

EXCURSIONS NEAR KIMMERIDGE BAY

Broad Bench towards Chapman's Pool

UPPER & LOWER KIMMERIDGE CLAY

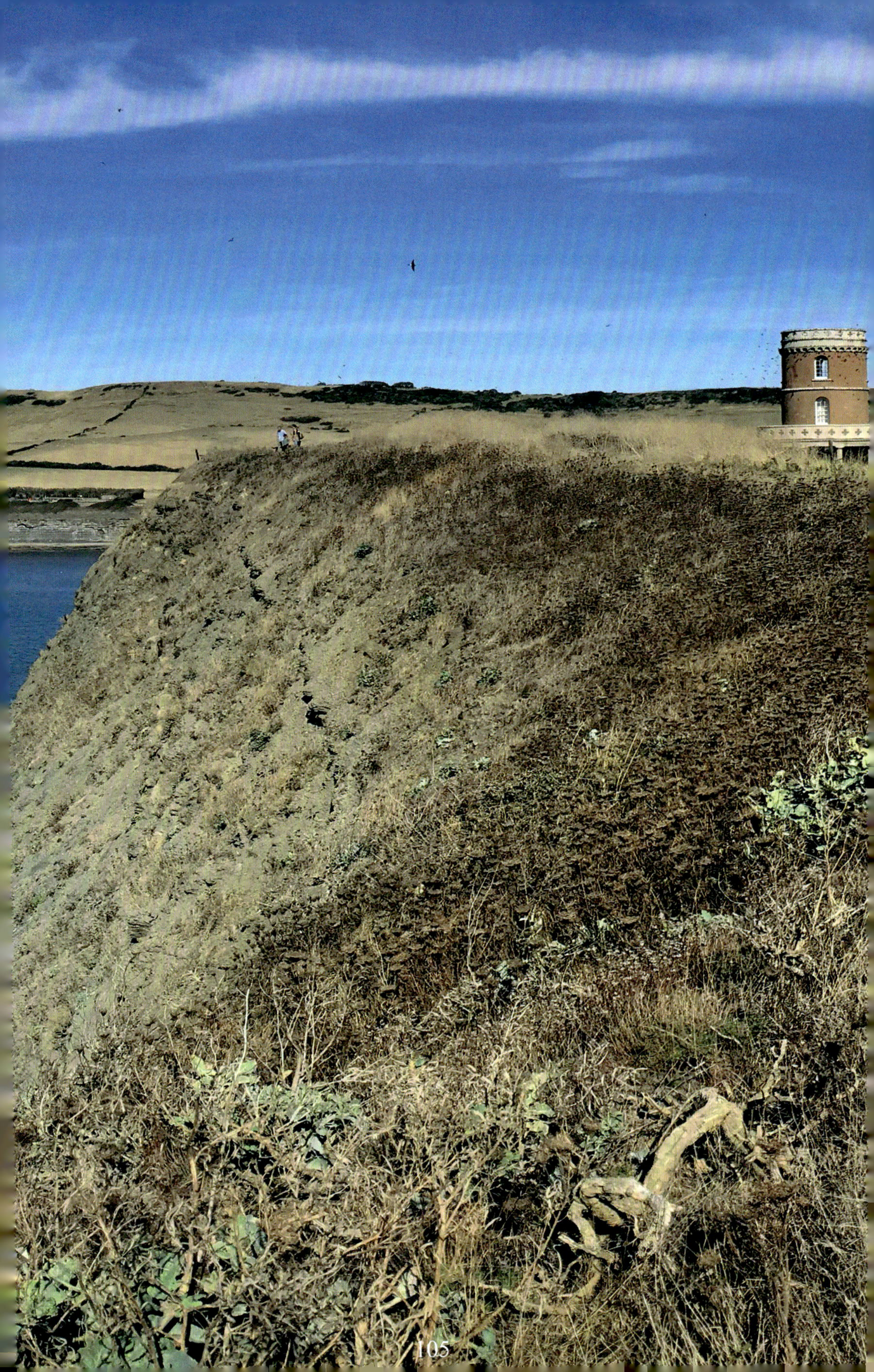

INTRODUCTION TO KIMMERIDGE BAY & SAFETY

The immense exposures of the Kimmeridge Clay Formation along the coast near the village of Kimmeridge are hugely important. Attaining a thickness of around 460 metres on the Dorset coast, it is one of the thickest sequences of mudrock in the UK. Within the rocks lies a wonderful fossil fauna that provides a vivid snapshot of the life and times in the seas, that would become Dorset, some 154.7–147.6 mya.

The areas accessible from Kimmeridge Bay are some of the most dangerous places to collect from anywhere along the Dorset coast. This very remote stretch of coastline is probably too dangerous for most people to visit. This is verified when viewed from above, on the south west coastal path. Here, the tides always reach the base of the tall, sheer cliffs and can quite easily cut you off. If you are intent on visiting the area to the east of Kimmeridge Bay, be aware this will involve extremely long walks around several headlands, so make it absolutely essential that you check and double check tide times. Be aware also of double tides. If you plan a walk from Kimmeridge Bay east towards Chapman's Pool, you might have little time to collect and you must reach your destination before the tide starts to turn, as it comes in very quickly. Please also be aware that east of Clavell Tower (Hen Cliff) there is no route up the cliffs.

Falling shale (often at high velocity) and potential cliff collapse presents another real hazard at Kimmeridge Bay and to the areas to the east and west. Hard hats should be worn at all times. Always keep away from the base of the cliff, to minimise any risk of accident from falling rocks and debris. In addition, the foreshore and ledges at Kimmeridge Bay are very slippery from significant algae and seaweed. Sensible footwear is essential.

To the far east of this excursion is Chapman's Pool, which lies on land owned by the Encombe Estate. The estate emphasise that fossil collection is *not permissible* here and for this reason, the location has not been included in this book.

Kimmeridge Bay, looking east, with Clavell's Tower in the distance, far right.

The collection of fossils at Kimmeridge Bay requires careful adherence to the conditions required by the landowners, Smedmore Estate. The use of geological hammers and chisels is absolutely forbidden and especially on the cliff faces and shale beds. Only loose fossil specimens (*ex situ*) may be collected from the rocks and ledges at Kimmeridge Bay. There are other localities where you may be able to use a hammer and chisel on loose rocks and nodules, however at Kimmeridge this is forbidden. This is a highly important SSSI site (Site of Special Scientific Interest).

Regretfully, the practice of hammering in the cliffs and bedrock is not unknown here. This totally unacceptable behaviour infringes on the wishes of the landowner and goes against both the JNCC and Dorset Fossil Collecting Codes and the rules of a SSSI. Please be aware that *only* loose fossils that can easily be picked up along the foreshore may be collected. The majority of fossils are largely crushed and the many ammonites seen out on the ledges are far better photographed as a record. In any case, they are far too fragile and compressed to attempt to collect.

Always tell others where you are and when you intend to return, but remember there is little, or no, mobile phone signal at Kimmeridge Bay.

Photographing fossils can provide a very good record of specimens found and can certainly provide a rewarding and suitable alternative to the physical collection of specimens, especially from such a sensitive location. This recommended safe practice, combined with a visit to the excellent Etches Collection Museum of Jurassic Marine Life, provides an opportunity to see and learn about the wealth of fossils collected at Kimmeridge Bay over the lifetime of one man; Steve Etches. The collection provides context and is a highly recommended visit for those interested in the fossils found in the area.

Ichthyosaurus sp. teeth. Image ©Etches Collection.

THE ETCHES COLLECTION
MUSEUM OF JURASSIC MARINE LIFE

Osteichthyes (bony fish). Image ©Etches Collection.

Image ©Etches Collection.

THE ETCHES COLLECTION
MUSEUM OF JURASSIC MARINE LIFE

Images ©Etches Collection.

A fine specimen of a Kimmeridgian fish, *Catrurus* sp. Image ©Etches Collection.

The Etches Collection is the result of one man's passion. Steve Etches has spent over 30 years amassing a collection of fossils from the area around Kimmeridge, the village in which he grew up and lives, which have now been housed in a modern purpose-built museum at Kimmeridge. His collection now consists of more than 2300 fossil specimens from the Kimmeridge Clay. Steve grew up and worked (as a plumber) in the area for 43 years but his passion for fossils has led to a world-class, unique and scientifically important collection.

Steve's collection is now housed in a permanent purpose-built museum at Kimmeridge – the Etches Collection: Museum of Jurassic Marine Life and is open to the public.
As a result of his achievement, Steve Etches was made a Member of the Order of the British Empire in 2014. In 2017, he was awarded an Honorary Doctorate by the University of Southampton.

An ichthyosaur. Image ©Etches Collection.

The Etches Collection is based in the village of Kimmeridge, eight miles south of Wareham in Dorset and sign posted from the A351 at Wareham. As you turn into the village, the museum building is on the right hand side, opposite Clavell's cafe and restaurant. Kimmeridge is close to Lulworth Army firing ranges and sometimes there can be road closures in the area. Visitors may need to take a detour via Corfe Castle or past Blue Pool to get to the museum at BH20 5PE.

Image ©Etches Collection.

KIMMERIDGE BAY

For this excursion, the starting point is from the Kimmeridge Bay cliff top car park, which is accessed via a narrow toll road, which passes through Kimmeridge village, continuing through Smedmore Estate. From here, access to the beach below is reached by descending a series of steps through a small valley at Gaulter Gap. The rocks are from the Kimmeridge Clay Formation, composed of fossil-rich mudstones and oil shales, which originally accumulated as soft sediment at the bottom of the sea approximately 156–148 million years ago. This interval of eight million years spans both the Kimmeridgian Stage and part of the subsequent Tithonian Stage of the Late Jurassic. At Gaulter Gap the cliffs are of Lower Kimmeridge Clay, comprising brown and hard bituminous shales and grey, softer mudstones of the *Aulacostephanus autissiodorensis* ammonite zone.

The Kimmeridge Clay outcrops locally in the cliff-face and on the foreshore between Brandy Bay in the west and Chapman's Pool towards the southeast. Fossils occur commonly throughout the clay, in particular the shells of ammonites (many very compressed) and bivalves. Less common finds include marine reptile remains and in extremely rare instances the bones of dinosaurs and pterosaurs. The formation comprises rhythmic alternations of soft mudstones, calcareous mudstones and kerogen-rich mudstones/shales. Layers of muddy dolomitic limestone occur at a number of levels and form prominent ledges which can be seen on the beach.

This pyritised fish was found by Viv Field and is a good example of taking a photo of a specimen found. In this case, extraction from the large slab would have proven too difficult and the fossil would have been destroyed in the process. In any case, the use of hammers is forbidden at Kimmeridge.

Once again, we must re-iterate that the use of geological hammers at Kimmeridge Bay is forbidden and any fossils found should be in loose, *ex situ* material and able to be picked up. The area beneath the car park is undoubtedly over-collected, especially in high season. The risk of rock falls is very high and in the search for fresh material, collectors will need to be vigilant and avoid the more dangerous areas, especially the base of cliffs. For those wishing to traverse sections of the coast beyond Kimmeridge Bay, it should be noted that this will not be possible on a rising tide.

Examples of the ammonite *Aulacostephanus pseudomutabilis* occur with some regularity in the Lower Kimmeridge Clay, which is exposed from about the centre of Kimmeridge Bay towards Broad Bench and again beyond to the west. You are able to walk some way westward in Kimmeridge Bay where you can study the Flats Stone Band. The western extremity of the bay though is in the Army firing ranges and there is no access along the beach west of the bay. There is, however, a cliff-top path which is open when the Range Walks are open (usually at weekends and certain holiday periods) and from which views to nearby Brandy Bay are possible.

Microconch of *Pectinatites* sp., an Upper Kimmeridge Clay ammonite at Kimmeridge Bay. Found and photographed by Viv Field.

The Flats at the western end of Kimmeridge Bay.

Steve Etches is the only collector allowed to extract and collect fossils from Kimmeridge Bay and its locality, due to his level of experience in collecting fossils from the Kimmeridge Bay strata for the sole purpose of conserving them for the benefit of science and education, and to be held in The Etches Collection museum.

Crushed ammonites are a common feature on the ledges of Kimmeridge Bay, all the way to Brandy Bay, to the west. Uncrushed specimens are rare and complete specimens are best seen at the Etches Collection.

The ledges here are typical of the Upper Kimmeridge Clay, with slippery, dark shale, with seaweed and blocks of dolomite. As Dr. Ian West has pointed out, "You walk over this dark shale, looking hopefully for a fossil bone; your chances of finding a bone are not very high. Steve Etches finds most!"

Marine reptile bone finds made by Bill Fagg, out on the ledges of Kimmeridge Bay. The large bone (above) is the scapula of a juvenile *Pliosaurus*, possibly *Pliosaurus brachyspondylus*. The vertebrae (below) are cervical vertebrae of a pliosaur.

The images on this page demonstrate the friable and delicate nature of fossils found in the shale ledges at Kimmeridge Bay. The fish (above) was found by Viv Field and the ammonites (below) were found and photographed by Bill Fagg.

Ammonite fossil in the rock platform in the western part of Kimmeridge Bay.

The Kimmeridgian Sea. Artwork by Andreas Kurpisz.

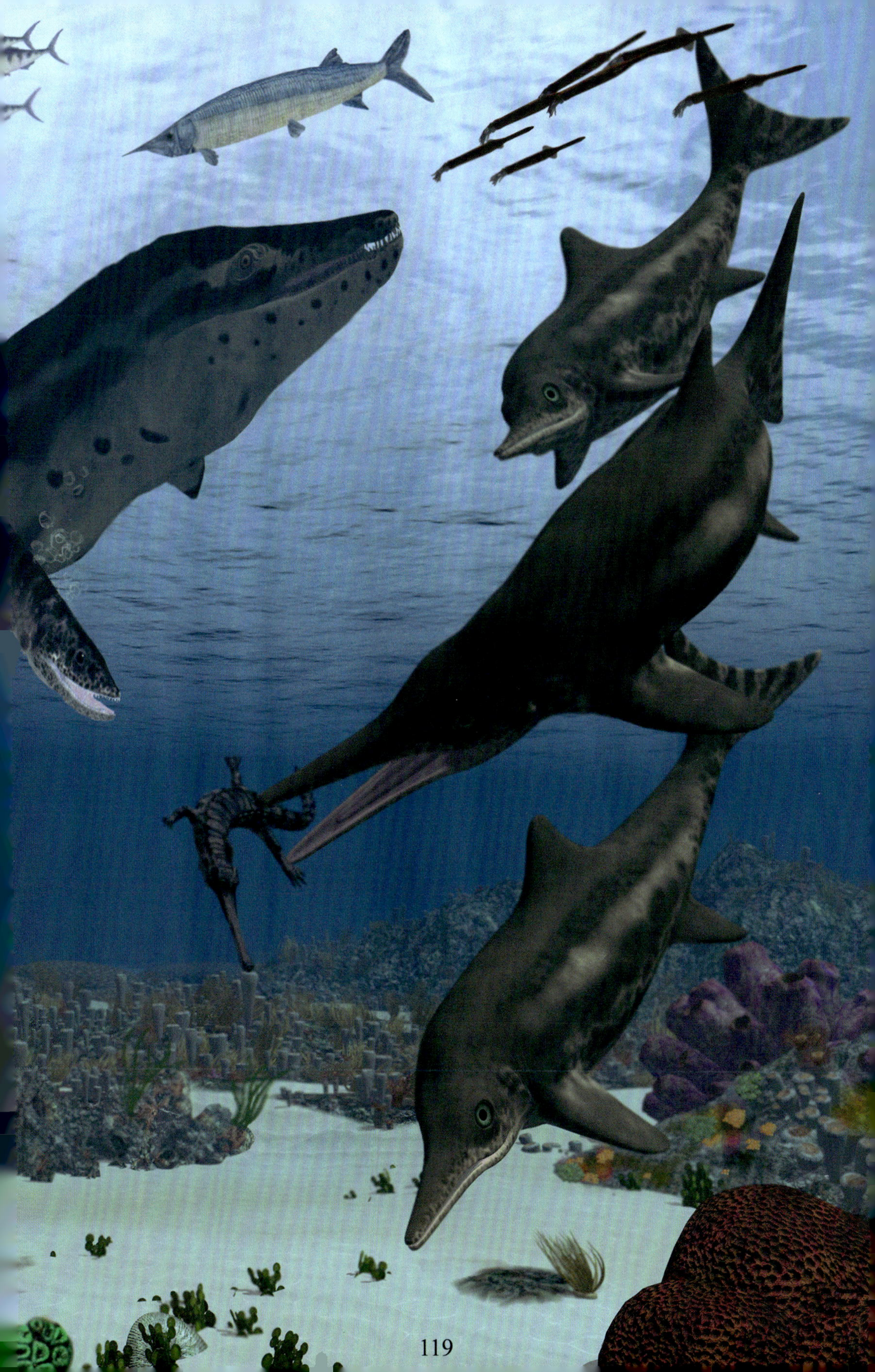

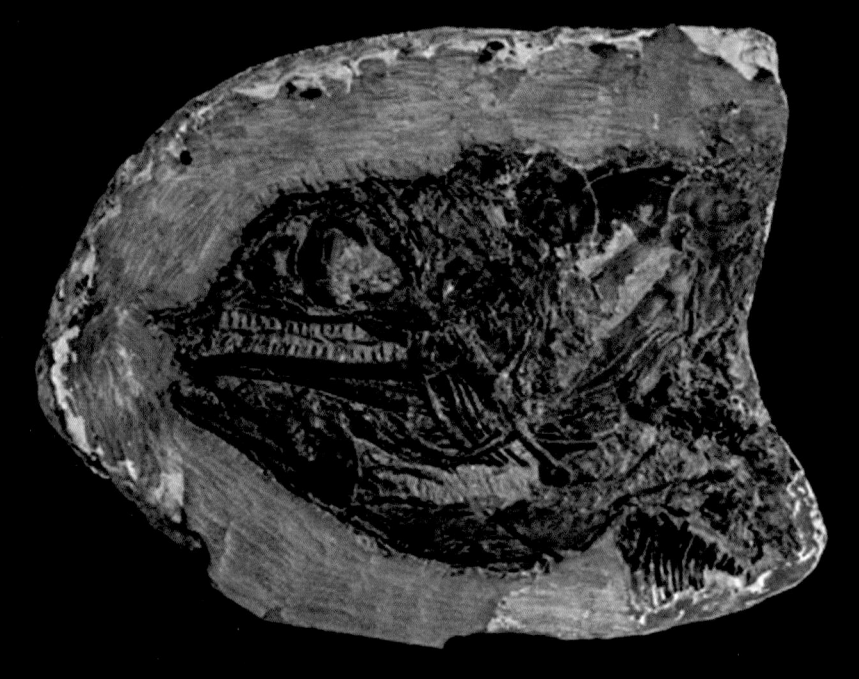

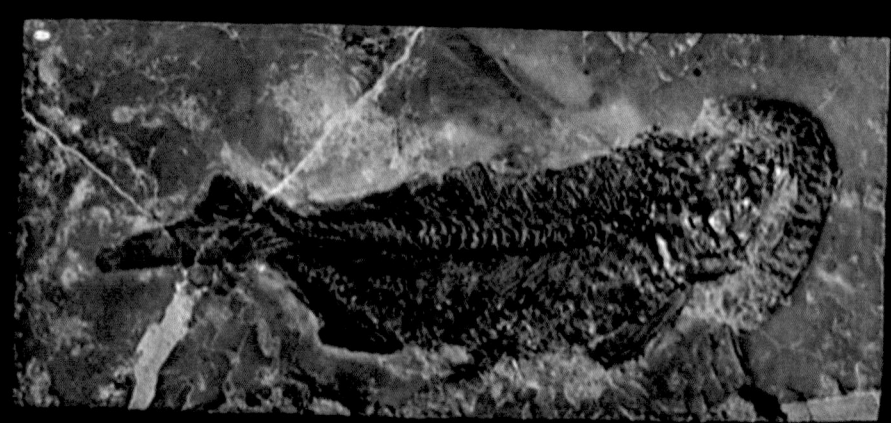

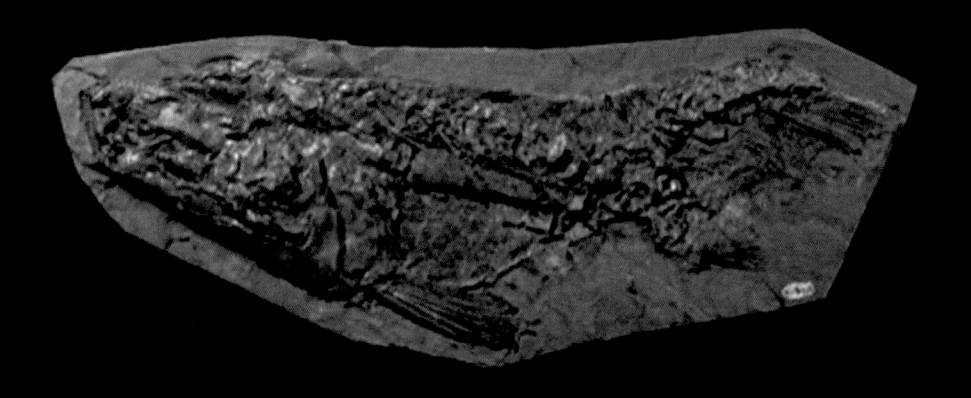

Various fossil fish held in the Etches Collection including (top to bottom): *Catrurus* sp., *Eurycormus* sp. and another *Eurycormus* sp. Images ©Etches Collection.

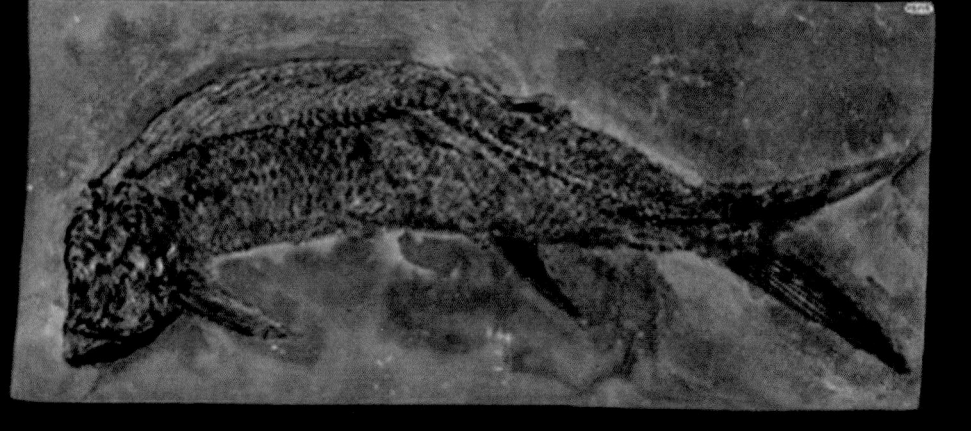

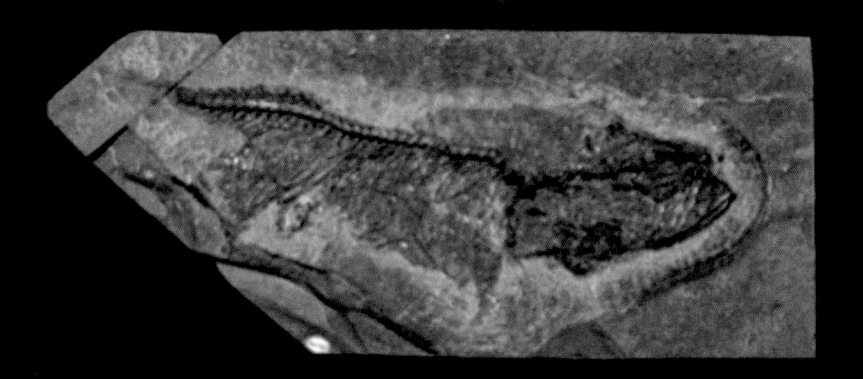

Various fossil fish held in the Etches Collection including (top to bottom): *Thrissops* sp., another *Thrissops* sp., and *Lepidotes* sp. Images ©Etches Collection.

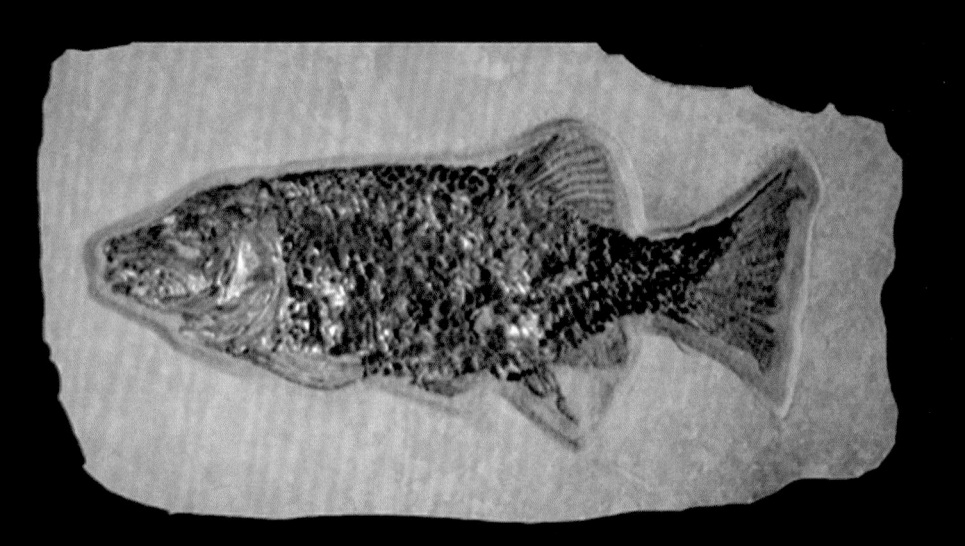

Plesiosuchus manselii, a genus of geosaurine metriorhynchid crocodyliform known from the Late Jurassic (late Kimmeridgian to early Tithonian stage).
Artwork by Andreas Kurpisz.

To the east of Kimmeridge Bay, the restrictions imposed by the Smedmore Estate still apply and visitors should again note that the estate does not allow the use of hammers. Permission to collect should also be obtained beforehand. This remote section must only be attempted on a falling tide. There are four tides a day here and the cliffs are sheer and crumbling, with a constant trickle of shale debris from above.

Ammonites immediately east of Kimmeridge Bay are almost entirely of the genus *Pectinatites* from the Bolonian Stage and usually crushed. In fact, the section eastwards shows the most complete and thickest section through the Upper Kimmeridge Clay in Britain and exposes the strata of the Bolonian Secondary Standard Stage in their entirety. Over 243 metres of Kimmeridge Clay are exposed along this section. Besides *Pectinatites*, the ammonite *Gravesia* also occurs in the Hen Cliff Shales beneath Hen Cliff. At the next bay, between Clavell's Hard and Rope Lake Head, a section through the *wheatleyensis* Zone shows very large pectinatitid ammonites but these are extremely fragile and cannot be removed. However, the small pyritised plates of the crinoid *Saccocoma* occur in these shales.

Access to sections further east is highly unlikely, as progress beyond the waterfall at Fresh Steps is impeded, unless there is an exceptionally low tide upon arrival. Even so, it can be daunting to potentially either wade in water or walk barefooted over extremely slippery rocks! This is one of the most challenging parts of the Dorset coastline and is not recommended for families or parties, or during winter and spring months.

Macroconch of *Aulacostephanus camericensis*, Lower Kimmeridge Clay, Kimmeridge Bay.

EXCURSIONS NEAR ST. ALDHELM'S HEAD

St. Aldhelm's Head & the Purbeck Quarries

PORTLAND GROUP
PURBECK GROUP

Coastal path approaching Tilly Whim Caves, near Durlston Head.

ST. ALDELM'S HEAD

The small outcrops of Kimmeridge Clay Formation below the west side of St. Aldhelm's Head are the last appearance of this rock along the East Dorset coast. At St. Aldhelm's Head, the Portland Group forms the summit of Emmett's Hill and St. Aldhelm's Head, capped by the lower part of the Purbeck Limestone Group.

The sheer cliffs of the south Purbeck coast offer spectacular coastal walking and the cliff-top path provides fine views along the coast, with occasional access to the sea, as at Winspit, Seacombe and Dancing Ledge. At St. Aldhelm's Head, a working quarry (St. Aldhelm's Head Quarry) occurs in the top of the Portland Stone and basal Purbeck Group, being the Lulworth Formation. At this quarry, the ammonite *Titanites anguiformis* occurs in the Shrimp Bed and even more abundantly in the underlying *Titanites* Bed. At the southern end of this quarry, the top surface of the Bed is exposed with numerous ammonite impressions, some well over 60 cm in diameter.

The pale beige Purbeck-Portland bed, the Spangle, is particularly rich in fossil shell content and in large *Titanites* ammonites, such as the one shown below. Haysom Purbeck Stone quarries this stone at its St. Aldhelm's Head site and often keeps a stock of these impressive fossils.

> Access to Haysom Quarry at St. Aldhelm's Head is by permission of the site owners. This is a privately owned, working quarry and the authors can take no responsibility for trespass or visits without prior approval from Haysom.

WINSPIT QUARRY

Winspit is an old disused quarry on the coastal cliffs near Worth Matravers. It is one of over 200 quarries that once existed around Worth Matravers, Langton Matravers and Swanage, which in their hey-day provided the stone for the prestigious buildings of London. Winspit was operational as a quarry right up until 1940.

Along with other quarries in the immediate area, it was a source of Portland Freestone from the Winspit Member. Specifically, the main stone extracted was Under Freestone, a hard, cream coloured oolitic freestone. The overlying bed, the Under Picking Cap (known locally as Spangle), was cut to waste in order to get to the Freestone beneath.

Fossils are not common, although impressions of the giant *Titanites* ammonite can be seen in many of the rock faces. Careful searching in the scree and loose material in the vicinity can still bring up the odd fossil from the Portland Formation.

Winspit Quarry, looking west. The fossil is of worm casts from the Dancing Ledge Member of the Portland Cherty Series.

Titanites anguiformis, a large perisphinctid ammonite. The shells of the ammonites were originally aragonite, lustrous, like that of a modern nautilus shell, and probably coloured. This was lost in solution after burial, probably when the Portland Stone was uplifted above sea level at the end of the Jurassic or in the early Cretaceous (i.e. Purbeck or Wealden times). Found at Winspit Quarry in the Portland Stone Formation and photographed by Viv Field.

Winspit Quarry from sea level. The rocks here are from the Portland Cherty Series (Dancing Ledge Member) which form the base of the quarry floor.

Whiteware Quarry, a Portland Stone quarry to the east of Dancing Ledge.

DANCING LEDGE & WHITEWARE QUARRIES

Dancing Ledge is a notable feature of the east coastal section of Dorset. This is a large gently sloping ledge of part of the Portland Cherty Series, with a large abandoned quarry above. It has been said to have received its name from the appearance of waves "dancing" on the ledge. The quarry, cliffs and ledge here provide a good section through the upper part of the Portland Stone Formation (Portland Cherty Series and through the Portland Freestone Member). The succession is similar to that found at Seacombe.

The Portland Stone that was once quarried from the cliffs at Whiteware Quarry is tens of metres above sea level. The next cliff quarry to the west is Dancing Ledge Quarry, which is much closer to sea-level as it is on the downthrow side of the Green Point fault. Both quarries ceased operation in the early 20th century and are now used for recreation by rock climbers.

The best fossils are those found within the rocks and boulders scattered on the quarry floor. Bivalves in the Portland Freestone are common but the sites have little else to offer.

Dancing Ledge, near Langton Matravers.

A reconstructed scene of the shores of a vast lagoon, in which a herd of brachiosaurs walked, leaving their footprints in the mud. Artwork by Andreas Kurpisz.

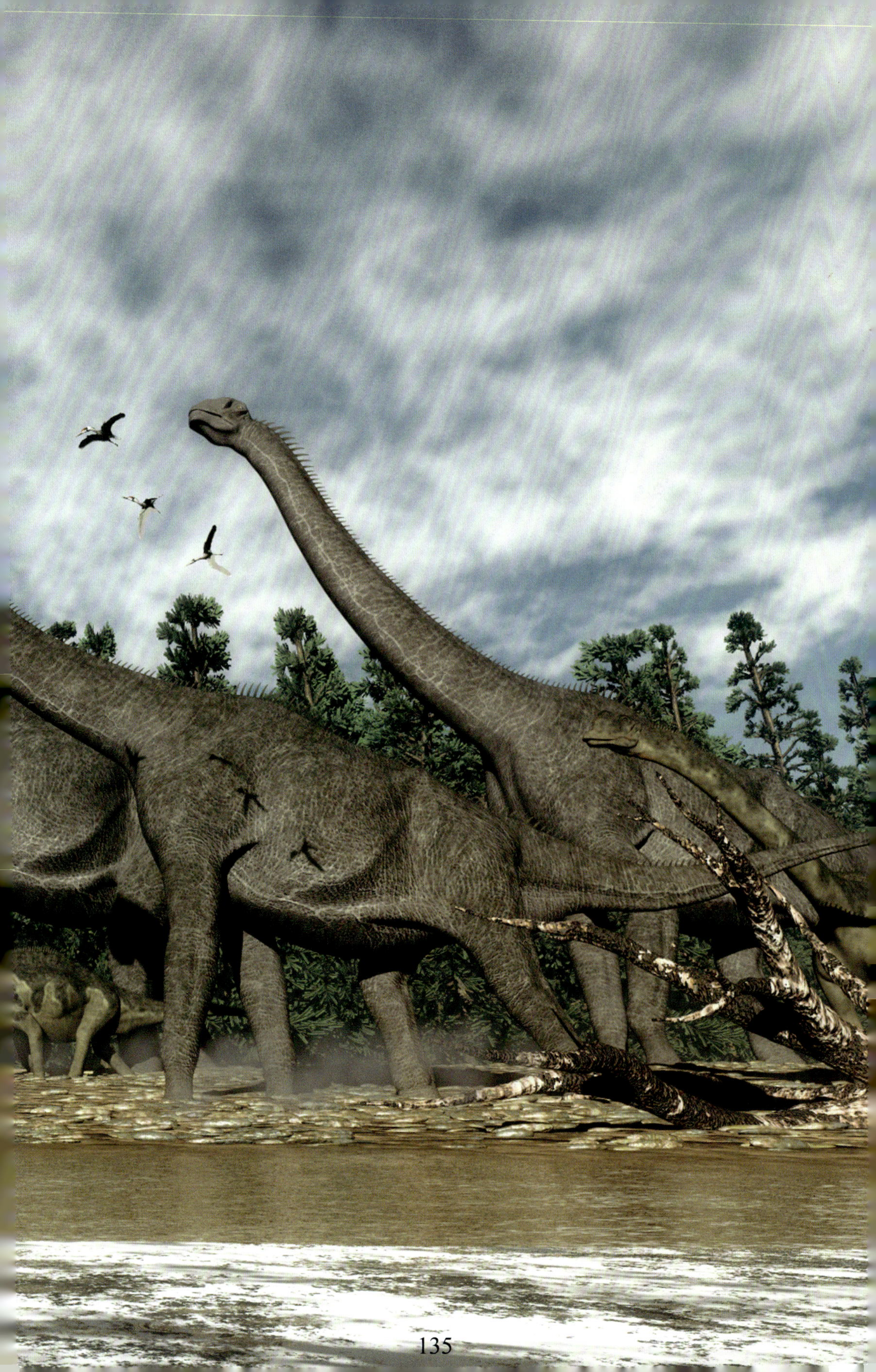

Footprint of a brachiosaur, from the trackway at Keates Quarry.

After the cessation of quarrying at the site, the prints were initially covered in fine gravel and after several years, long term responsibility for the site was handed to the National Trust, who own the land in which the quarry sits. The land around it has since been secured for safe public access. Access to the trackway is *not* gained through the quarry but from Spyway car park at BH19 3HG. From Spyway, follow the footpath south towards the sea and turn right onto the Priest's Way, heading west. Continue on the Priest's Way for about three quarters of a mile – the dinosaur footprints are signposted on the right, just past the turning for Acton.

The Purbeck limestone in this area can be highly fossiliferous and the whole area along Priest's Way is littered with spoil from the many working quarries and can provide some interesting finds with a bit of persistence. *Unio*, the freshwater bivalve is a common find. However, entering into the various working quarry sites is not permissible, unless the site owners have been contacted prior to any visit and agreed.

The fossils here provide evidence of a large Cretaceous lagoon, with fish, plants, freshwater bivalves, assemblages of teeth and bones of fish and the shark *Hybodus*, turtle bones and carapace fragments. Rare small mammal bones and teeth are also frequently found.

Turtle carapace and fish jaw found by Martin Curtis.

A block containing bones and scales of a fish, probably *Lepidotes* sp. (now *Callipurbeckia minor*) found by Nicola Parslow.

Archaeoniscus broiei, a crustacean related to a modern wood louse. Found by Nicola Parslow.

Macromesodon daviesi, a pycnodontid or bony fish. Found by Kevin Keates Snr.

Keates Quarry, near Worth Matravers.

LEWIS QUARRY

Lewis Quarry in Acton is another private, working quarry supplying Purbeck Stone for commercial purposes. It is not publicly accessible.

In the summer of 2018, it was announced that a set of sauropod prints had been found, which are considered to be of the same herd that were discovered at adjacent Keates Quarry in Worth Matravers. The series of parallel prints are distinctly 'saucer-shaped,' although shallower than those at Keates. The dip of the beds, which folded when the European Alps were pushed up, means that the tracks are closer to the ground in Keates Quarry where they can be preserved but are much deeper at Lewis Quarries where *in situ* preservation is not possible.

> Access to Lewis Quarry at Acton is by prior permission of the site owners only.
> This is a privately owned, working quarry and the authors take no responsibility for trespass or visits without prior approval.

Lewis Quarry and the work being undertaken to expose and record the sauropod trackway.

CALIFORNIA QUARRY & PANORAMA QUARRY

These two quarries are in close proximity and are owned, by J. Suttle (California Quarry) and Haysom (Panorama Quarry). Both extract Purbeck Stone but because of deeper extraction than at Keates Quarry, the beds of the lower part of the Intermarine Member are exposed. At these quarries, dinosaur footprints are found with some regularity and occur in strata that have low palaeosalinity, meaning the truly fresh water was drinkable for the reptiles that frequented the area. The tracks here are tridactyl and occur just below the Roach layer. This is a bed of middle Purbeck Limestone (i.e. within the Stair Hole Member [18–40 metres] of the Durlston Formation).

In 1961, a trackway was discovered at Suttle's California Quarry. The trackway was unusual in that it was the longest trackway that had been found in the UK. It was identified as being the print of *Iguanodon* and at first it was thought to be the trackway of one animal, but closer research revealed that it was in fact two sets of dinosaur prints; *Iguanodon* and *Megalosaurus*. Part of the trackway can now be found in the garden at the Natural History Museum, London.

Access to California Quarry and Panorama Quarry is only by prior permission of the site owners. These are privately owned, working quarries and the authors take no responsibility for trespass or visits without prior approval.

California Quarry, Swanage.

Formation	Member	Beds		Details
DURLSTON FORMATION	Peveril Point Member			Blue Marble
				Grey Marble
				Green Marble
				Broken Shell Limestone
	Stair Hole Member	Intermarine Beds or Upper Building Stones		Thin Limestones & Shales
				Laning Vein
				Thin Limestones & Shales
			Freestone Vein	Red Rag
				Under Rag
				Grub
				Roach
				Thornback
				Whetston
				Freestone
				Blue Bed
				Thin Limestones & Shales
				Downs Vein
				Cinder Bed
LULWORTH FORMATION	Worbarrow Tout Member			New Vein — Cherty Freshwater Beds
				Many beds of thin Limestones & Shales
				Cypris Freestones
	Mupe Member			Thin Limestones & Shales Broken Beds
				Portland Freestone Member (Portland Stone Formation)

SIMPLIFIED SECTION TO ILLUSTRATE THE WHOLE SEQUENCE OF PURBECK BUILDING STONES
Not to scale.

Dinosaur print found by Lizzie Hingley.

The spoil heaps at the quarry site.

The main quarry face.

Large footprint of an iguanodont at Panorama Quarry; one of many tridactyl prints found here. The footprints occur in low palaeosalinity strata. The location is at the deepest part of the basin (i.e. the thickest Purbeck sequence) in which the lakewater was drinkable.

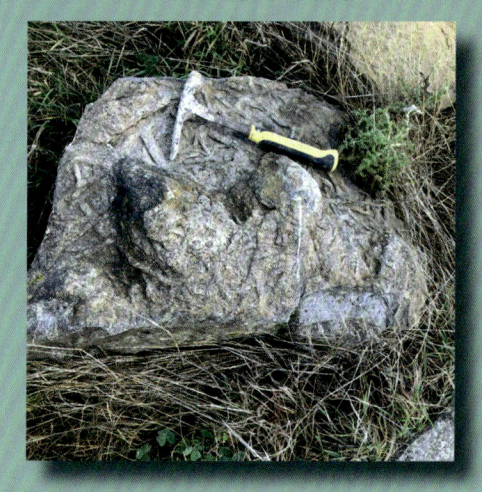

The fish are *Callipurbeckia minor* (formerly *Lepidotes minor*) and renamed by Adriana Lopez-Arbarello (2016). Found in the Stair Hole Member, Intermarine Bed of the Durlston Formation. Both specimens found by Steve Snowball and Craig Chivers and prepped by Craig Chivers.

EXCURSIONS NEAR DURLSTON BAY

Durlston Bay, Peveril Point & Swanage

PURBECK GROUP
WEALDEN GROUP
UPPER GREENSAND FORMATION
CHALK GROUP

Swanage Bay, with Durlston Bay in the foreground.

DURLSTON BAY

> **SAFETY NOTE**
> This section can be quite hazardous, as the cliffs are unstable, access to the beach is difficult and the very rocky foreshore is covered in algae and seaweed. Please take note of tide times prior to examining the section, as this location can have unpredictable tide conditions. Double tides frequently occur and the sea here can be very powerful.

The two kilometre stretch of coastal sea cliffs between Peveril Point and Durlston Head near Swanage displays the finest sections of the Purbeck Limestone Group in Britain. Durlston Bay is the type locality for the Purbeck Limestone Group. Here, the base of the Durlston Formation, below the Cinder Bed, can be observed. Much of the complete succession is obscured by rock falls, landslips and a coastal protection scheme, built to safeguard the cliff-top flats in 1989. However, Durlston Bay is a fascinating place, although possibly better suited for the geologist or experienced fossil collector.

Dinosaur print in the Durlston Formation at Durlston Bay. This is undoubtedly the footprint of a sauropod dinosaur from the Intermarine Beds. Found and photographed by Viv Field.

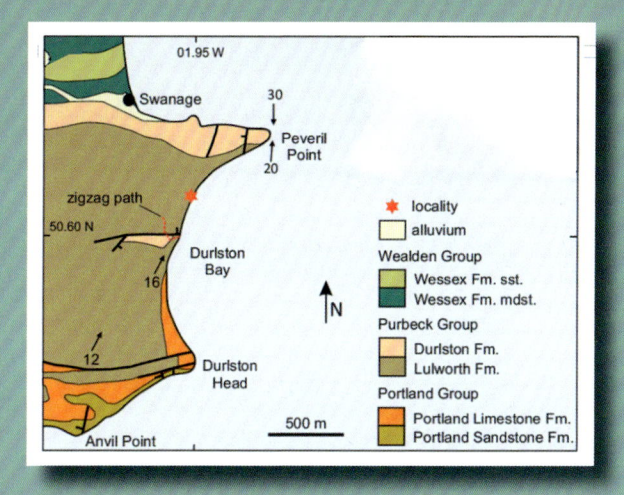

Durlston Bay is one of the most important Early Cretaceous fossil sites, known for the remains of turtles, early lizards, dinosaurs, crocodiles, pterosaurs, early mammals and insects that lived in the swampy, lagoonal ecosystem that existed 140–145 mya.

Durlston Bay represents a transition between the Jurassic and the Cretaceous periods. During this time a large lagoon stretched along the coast and as a result, the most commonly found fossils are teeth belonging to fish and crocodiles (although the latter are significantly less abundant), bivalves, fragments of turtle carapace and bone, crocodile scutes and isolated fish bones.

Cliff section of the area near to Peveril Point (adapted from Ian West). Image used with kind permission of Roy Shepherd at Discovering Fossils.

Owenodon is a genus of iguanodontian dinosaur known from a partial lower jaw discovered in Early Cretaceous-age rocks of Durlston Bay. Also depicted are several *Echinodon*, a genus of heterodontosaurid dinosaurs that also lived in the area during Berriasain times, with remains also found in the Lulworth Formation, Worbarrow Tout Member. A pair of *Plataleorhynchus streptophorodon* pterosaurs fly overhead. These are also known from the Purbeck Limestone and with teeth found in a chalkstone quarry near Langton Matravers.
Artwork by Andreas Kurpisz.

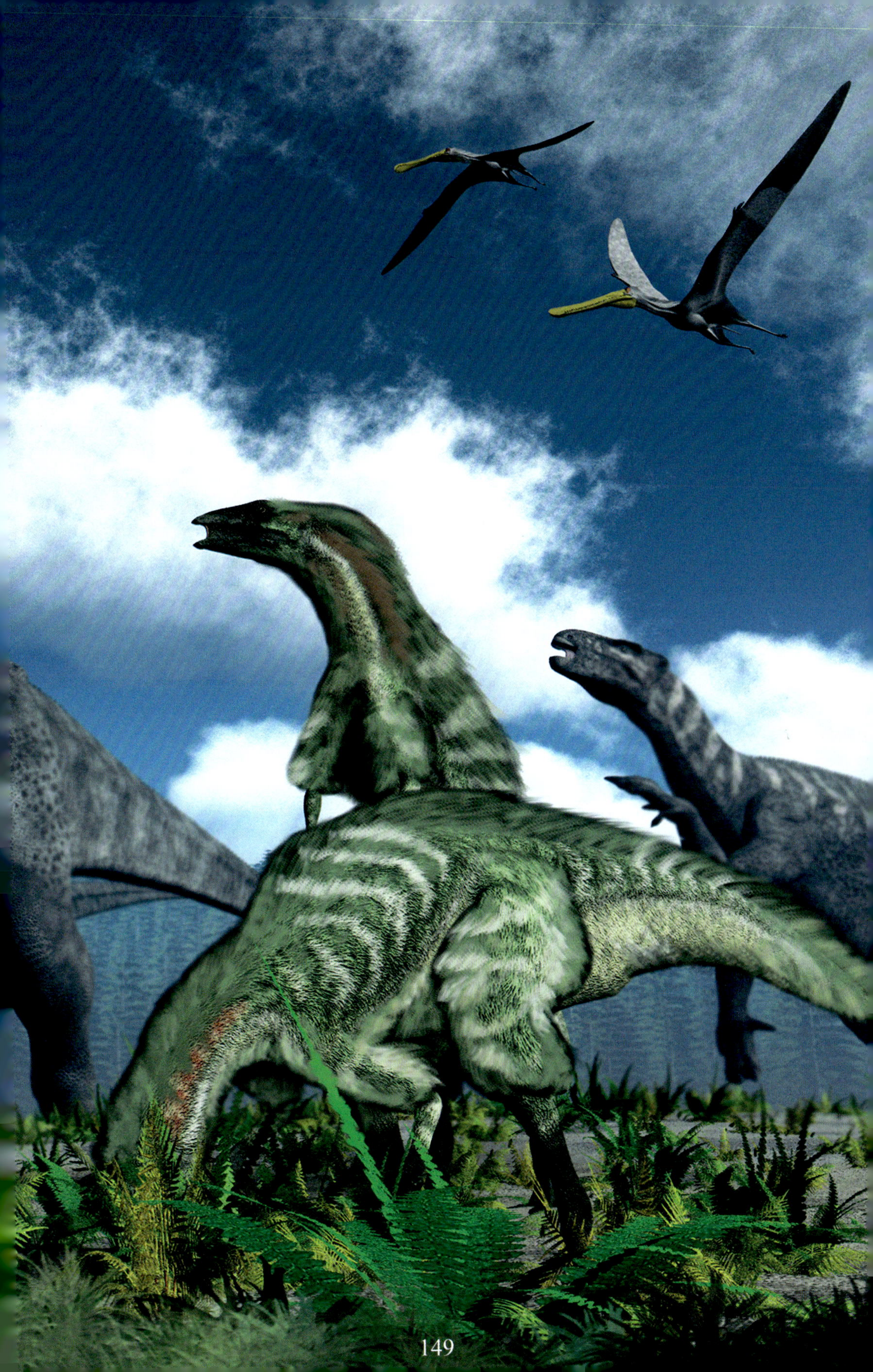

Formation	Member	Bed	Thickness
DURLSTON FORMATION	PEVERIL POINT MEMBER	OSTRACOD SHALES MEMBER	11.5 m
		UNIO BEDS	1.1 m
		BROKEN SHELL LIMESTONE	2.9 m
	STAIR HOLE MEMBER	CHIEF BEEF BEDS	8.2 m
		CORBULA BEDS	11.1 m
		SCALLOP BEDS	1.6 m
		INTERMARINE BEDS	15.6 m
		CINDER BED	2.95 m
LULWORTH FORMATION	WORBARROW TOUT MEMBER	CHERTY FRESHWATER BEDS	8.1 m
		MARLY FRESHWATER BEDS	4.2 m
		SOFT COCKLE BED	22.0 m
		HARD COCKLE BED	4.0 m
	RIDGEWAY & MUPE MEMBERS	CYPRIS FREESTONES	19.4 m
		BROKEN BEDS & CAPS	8.9 m

The succession in Durlston Bay (modified after Clements, 1993). Not to scale.

DURLSTON FORMATION

PEVERIL POINT MEMBER

At Peveril Point, the uppermost Durlston Formation of the Purbeck Group, belonging to the Lower Cretaceous, is seen. Access to the Durlston Formation is easy from Peveril Point but on the Durlston Bay side of this headland, access along the beach is often restricted.

Here, the Cinder Ledge and Cinder Bed form the reef and basal bed, but falling rock from the cliffs may well inhibit progress past the Cinder Bed. There is some real risk near Peveril Point, but there is a greater main hazard area further south where the cliffs are higher. Here, the Broken Shell Limestone may fracture and fall to the shore. This is fairly common and can be quite dangerous in places. Safety helmets are advised (except at Peveril Point). In inclement weather, it is probably wiser for parties not to proceed far south of Peveril Point via the beach. The cliff section further south and in the Middle Purbeck can be ascended at the Zigzag Path.

Durlston Bay, towards Peveril Point.

DURLSTON FORMATION – PEVERIL POINT MEMBER

Peveril Point provides the only location for 'Upper Purbeck' strata and is made up of the uppermost three beds of the **Peveril Point Member** (see chart on page 150). Here, the **Upper Ostracod Shales Member** (formerly the Upper *Cypris* Clays and Shales) are found. These grey shales, often darkened by pyrite, contain the ostracod, *Cypridea* and numerous examples of the pond snail *Viviparus*. These animals thrived on luxuriant aquatic vegetation in lacustrine (lake) environments.

In the underlying **Unio Beds,** the freshwater mussel *Unio porrectus* is also common, in a thin green, sandy limestone found here, along with teeth and scales of fish (*Pycnodus, Lepidotus*), which are shiny and dark.

The **Broken Shell Limestone** is known as 'the Burr' – a massive, light creamy-grey biosparite, containing broken shells, as its name suggests and dominated by shells of *Viviparus, Unio* and *Neomiodon*, with fragmentary turtle and fish remains.

Above left: *Coelodus* sp. fish teeth. **Bottom left:** Fish scale, probably *Callipurbeckia* (*Lepidotes*) sp. **Above right:** Assortment of fish teeth. Found by Viv Field in the Unio Beds, Peveril Point.

DURLSTON FORMATION – STAIR HOLE MEMBER

The bottom-most bed in the succession of the Stair Hole Member is the **Cinder Bed,** a bluish-grey oyster bed in the northern part of the bay. The bed is so packed full of shells it resembles a hardened bed of cinders, forming a projecting ledge. The shells are those of the lagoonal oyster *Praeexogyra distorta,* and the location can be easily reached by proceeding along the shore from Peveril Point. Due to faulting, there is another shore ledge exposure in the northern part of the bay and also in the cliff immediately to the north of the Zig Zag path, in the middle part of the bay.

The Cinder Bed ledge exposed at the northern end of Durlston Bay and a section of the bed packed with the oyster *Praeexogyra distorta* (inset).

Goniopholis simus
Lower Cretaceous (Berriasian) crocodile,
Intermarine Bed, Stair Hole Member,
Durlston Bay.
Found by Richard Edmonds.

Goniopholis kiplingi in a scene with Metriohynchus, in late Jurassic seas, in the area that is now around Swanage. Artwork by Andreas Kurpisz.

The Stair Hole Member at Durlston Bay is almost entirely made up of the **Intermarine Beds**, known locally as the Upper Building Stones and which forms the 'Middle Purbeck' of old texts. This section is reached via the Zig Zag path (see page 157), but take note that this is one of the main areas specifically to avoid when conditions are unsafe. South of the Cinder Bed Ledge in the northern part of the bay, the lower part of the cliff is vertical with loose, falling debris from above. This is certainly a section to avoid in late winter and early spring, particularly after or during heavy rainfall.

The Intermarine Beds consists of a number of beds, which represent different conditions during deposition. The Red Rag Bed is interesting as this medium-grey rock, mottled with creamy brown is a source of many fossil shells, notably *Neomiodon,* which is well preserved in the shale beds, along with other bivalves such as *Unio, Praeexogyra, Isognomon, Chlamys,* etc. The top surface can yield reptile bones, and plant debris including leaves. Gastropods include *Viviparus* and *Ptychostylus harpaeformis.* The skull of the crocodile, *Goniopholis simus* was found in 2007, not far below the horizon of the Red Rag Bed and might well represent the remains of a drifting carcass.

This skull of the Middle Purbeckian crocodile *Goniopholis simus* was found by Richard Edmonds (centre) at Durlston Bay and extracted with the help of Chris Moore (left), his son Alex Moore and Steve Etches (right).

To access the middle section of Durlston Bay, take the Zig Zag path. You will need to park down Belle Vue Road BH19 2HP, just off of Durlston Road and take the footpath that is signposted as the 'Zig-Zag' path, which goes between the flats. The Zig-Zag path will take you directly onto the foreshore.

Turtle bone
Intermarine Bed, Stair Hole Member,
Durlston Formation, Swanage.
Found by Anastasia Jeune.

Intermarine Bed, Stair Hole Member, at Durlston Bay.

In situ long-toed theropod print in the Durlston Formation, Intermarine Beds of Durlston Bay. Photo by Viv Field.

DURLSTON BAY DINOSAURS

The tracks of dinosaurs are a regular occurrence in both the central and southern parts of Durlston Bay, within the Intermarine Beds, particularly the Roach, a beach deposit well known as the main dinosaur footprint bed. This lagoon beach surface probably remained exposed for a long period, which would explain the reddening of the top.

Sauropod print in the Roach of Durlston Bay. Photo by Viv Field.

Iguanodon footcast
Intermarine Beds,
Purbeck Group.
Image used with kind permission
of Swanage Museum.

Turtle carapace from the Intermarine Beds of the Stair Hole Member. Turtle shell is common in the Purbeck Beds and is very distinctive as it occurs in broad, flat plates. The bone structure is also distinctive: typically, very open with large, calcite-filled voids between the darker brown bone. Found and prepared by Craig Chivers.

Three further beds make up the remainder of the Stair Hole Member. The thin **Scallop Beds** lie directly above the Intermarine Beds, with the *Corbula* **Beds** above that and finally the **Chief Beef Beds**. If approaching from Peveril Point, the succession will be from the top downwards (see page 150 for the succession of the four members at Durlston Bay). The Scallop Beds yield pectin bivalves. The *Corbula* Beds are thicker and where the most common fossils are bivalves of *Corbula* and *Protocardia* and ornamented gastropods. The Chief Beef Beds consist of paper shales, with layers of 'beef' fibrous calcite and a high organic content along with shelly limestones, where the main fossil is the bivalve *Neomiodon*.

Purbeck limestone from the *Corbula* Beds, with bivalves of *Neomiodon medius* and *Corbula alata*.

LULWORTH FORMATION – WORBARROW TOUT MEMBER

In 1857, following the discovery of a mammal jaw at Durlston Bay, ex-lawyer and fossil collector Samuel Beckles was persuaded to direct a major excavation in the area that the jaw was found. This became known as 'Beckles' Pit,' with five metres of overburden removed over a 600 square metre area. It was one of the largest scientific excavations, ever. The Mammal Bed, a layer of marl known as a 'dirt bed' averaged only 13 cm in thickness. The excavation lasted for nine months and was featured in the Illustrated London News in December 1857 (see illustration below). As a result of the excavation, at least twelve species of mammals were recovered, as tiny fragments of jaw and teeth, along with the remains of reptiles, insects and freshwater shells. Following cessation of Beckles' quarrying activities, very few mammal specimens have been reported from this horizon and the quarry, which is in the upper part of the cliff, is now obscured and inaccessible.

The Mammal Bed of Beckles lies beneath the Cinder Bed, in the **Marly Freshwater Beds**, with occurrences of other mammals in the underlying **Cherty Freshwater Beds**. However, recently, residue obtained from screen washing small samples taken from the Mammal Bed and adjoining beds, exposed at beach level in Durlston Bay, have continued to yield isolated mammal remains.

SAFETY NOTE
The area, to the south of the northern Cinder Bed Ledge has been subject to frequent rock-falls and the risk still exists. The cliff was undercut at the Mammal Bed and an open fissure has developed above. The section is certainly best studied at low tide, although the tidal range is not great and if cut off, ascent is by the Zigzag Path. Do not attempt an ascent of the cliffs where there is no well-defined path. At low tide beware of slipping on slimy rocks, especially on the Upper Building Stones north of the Cinder Bed Ledge. The ledges nearer to Durlston Head may be swept by waves.

Towards the middle section of Durlston Bay, the Mammal Bed reaches the foreshore. Note that the beds suddenly dip in the middle of Durlston Bay and run for a while along the foreshore. This is the point at which the Mammal Bed reaches the level of the beach. The bed can be seen sandwiched between two hard limestone layers and is comprised of a thin black layer, followed by a grey layer, then a very thick second black layer, with a lighter grey layer resting on top of the lower limestone. Mammal remains can sometimes be found within the harder limestone but these are rare. It is much easier to search through the shale. The Mammal Bed can easily be recognised by the colour and the number of freshwater shells within it.

Fossil enthusiast, Ian Cruickshanks has meticulously collected micro-fossils and mammal remains at Durlston Bay over many years. His collection is currently being studied at the University of Portsmouth.

Within the dark banded freshwater Mammal Bed, fragments of bone are sometimes found, which are compacted within the soft black and grey layers. Fragments of crocodile, fish and turtle are also found here. However, the bones are *extremely* fragile, so it is best to leave them within the shale and then extract and preserve them when at home. An ideal way to collect here is to take several bags of samples home. The samples should be thoroughly air dried before being processed in small batches.

Processing involves disaggregation in fresh water and hydrogen peroxide. Samples can then be hand sieved in water using a sieve with a 0.5 mm mesh. Ensure that you sieve at 0.5 mm, since many of the bones or teeth will fall through anything wider than this size of mesh. Dried residues can then be picked under a binocular microscope or strong hand lens. Mammal teeth recovered are usually between 0.5–2.0 mm in size.

The Purbeck Group Mammal Bed, was, until the latter part of the twentieth century, one of very few horizons world-wide to yield Mesozoic mammals. Recently, residue obtained from screen washing small samples taken from the Mammal Bed exposed at beach level have yielded isolated mammal teeth, including the two upper molars discussed in detail on page 164.

Cliff at the middle of Durlston Bay, approximately at the site of Beckles Mammal Bed. This section of coast is very unstable. Approach with extreme caution.

In 2017, around 55 kg of bulk sampling material was removed from the location at Durlston Bay and in the first instance, thoroughly air-dried in the laboratory, before being processed in small batches. Dried residue was divided into size fractions, using a nest of sieves and from which mammal teeth were recovered from particle sizes of 0.5–2.0 mm. Two indeterminable mammal teeth were recovered from one of the samples, along with a very small incisor. A double-rooted canine and the cusp of a crown were recovered from other samples.

The two teeth are of huge scientific importance and have now been shown to be those of two different eutherian mammals (i.e., those mammals lacking an epipubic bone, thus allowing for the expansion of the abdomen during pregnancy). They have been named *Durlstotherium newmani* and *Durlstonodon ensomi*. The beach-level exposure from which the new teeth were recovered is of variable extent, due to changing beach conditions but whilst not unknown, beach level discoveries such as this are still relatively rare.

Further research conducted by Prof. Steven Sweetman, Grant Smith and Prof. David Martill, along with funding for the project at the University of Portsmouth, conclude that *Durlstotherium* and *Durlstodon* represent the earliest eutherians currently recorded in Europe and the second oldest placental mammals ever found. The importance of these finds made by Grant Smith is immense. These mammals were right at the origin of the group of mammals to which we, as humans, belong. The small mammals would have lived alongside the dinosaurs and were most probably nocturnal in habit, with *Durlstotherium newmani* being a possible burrower that ate insects, whilst *Durlstonodon ensomi* may have eaten plants as well. The teeth are of a highly advanced type that can pierce, cut and crush food. They are also very worn which suggests the animals to which they belonged lived to a good age for their species.

Two fossil teeth of the Purbeck Mesozoic mammals *Durlstotherium* (A1–4) and *Durlstodon* (B1–4). These Jurassic mammals are ancestors to placental mammals and are the oldest of their kind found in Europe. Photograph reproduced from Sweetman *et al.* (2017).

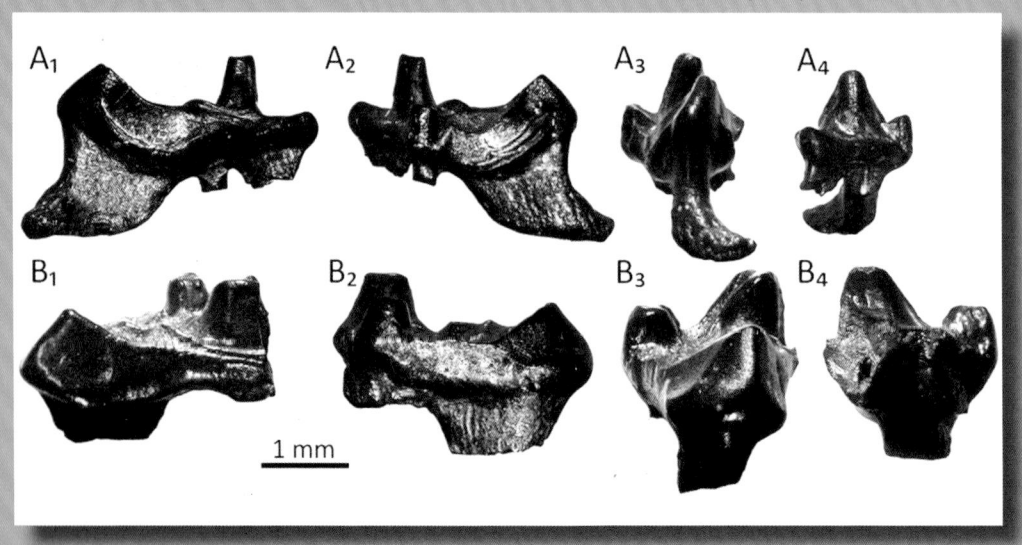

Artist's impression of the Purbeck lagoon at dusk with *Durlstodon* (left foreground), *Durlstotherium* (right and centre foreground) and the theropod *Nuthetes* holding a captured *Durlstotherium* (centre middle distance). Artwork ©Mark Witton and used with his kind permission, reproduced from Sweetman *et al.* (2017).

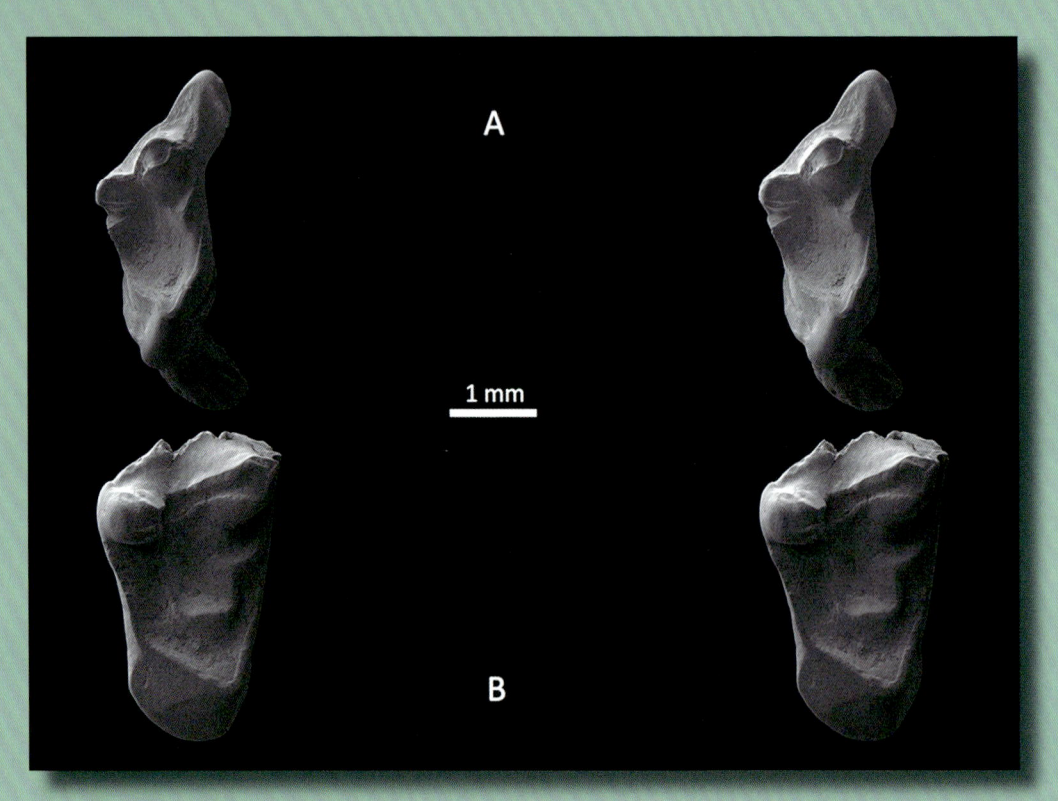

Stereo scanning electron micrographs of studied eutherian mammal specimens from the Berriasian Purbeck Group of Dorset, southern England; in occlusal view. A. *Durlstotherium newmani*, NHMUK PV M 99991. B. *Durlstodon ensomi*, NHMUK PV M 99992. Images used courtesy of Dr. Steve Sweetman and reproduced from Sweetman *et al.* (2017).

Opposite page: Stratigraphic log of part of the Purbeck Group exposed in the northern part of Durlston Bay showing the horizons from which mammal remains have been recovered, and the horizon from which the specimens illustrated here were obtained (modified from Sweetman *et al.*, 2017).

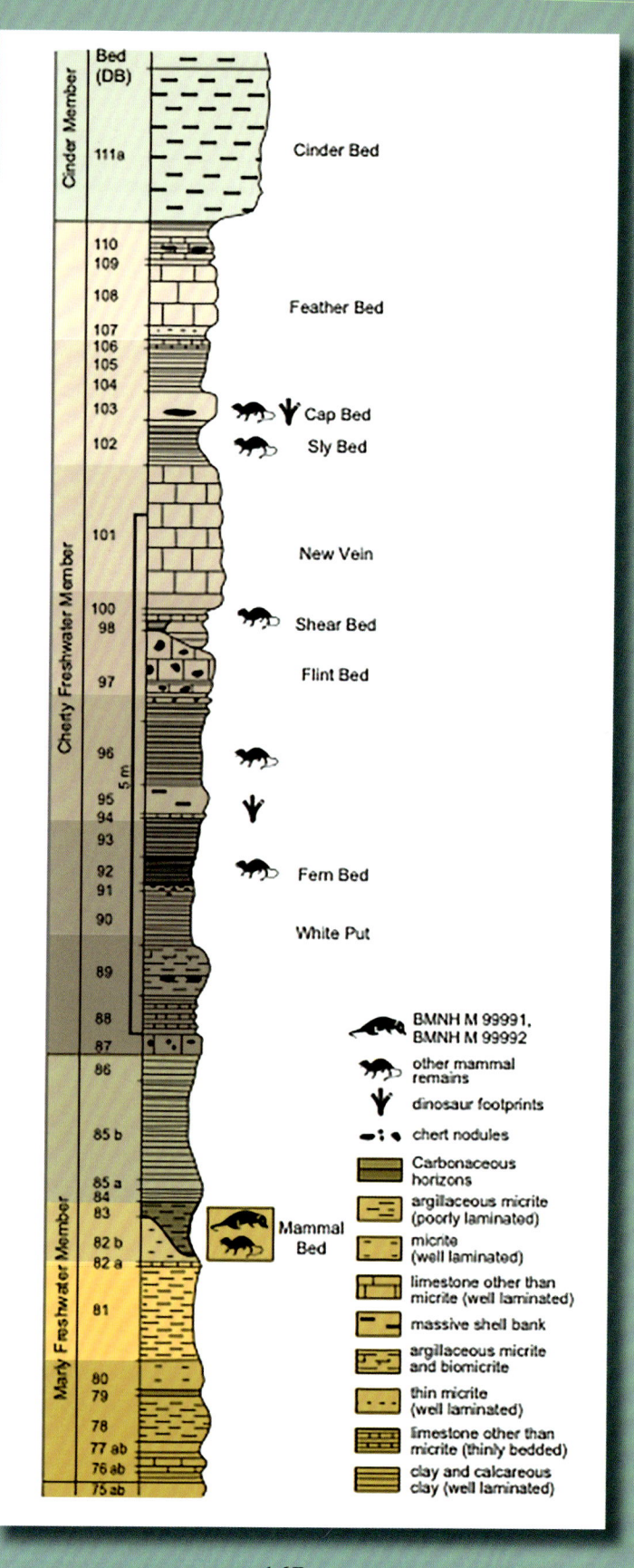

Member	Bed (DB)		
Cinder Member	111a		Cinder Bed
Cherty Freshwater Member	110, 109		
	108		Feather Bed
	107, 106, 105, 104		
	103		Cap Bed
	102		Sly Bed
	101		New Vein
	100, 98		Shear Bed
	97		Flint Bed
	96		
	95, 94		
	93		
	92, 91		Fern Bed
	90		White Put
	89		
	88		
	87		
Marly Freshwater Member	86		
	85 b		
	85 a, 84		
	83		Mammal Bed
	82 b, 82 a		
	81		
	80, 79		
	78		
	77 ab		
	76 ab		
	75 ab		

Legend:
- BMNH M 99991, BMNH M 99992
- other mammal remains
- dinosaur footprints
- chert nodules
- Carbonaceous horizons
- argillaceous micrite (poorly laminated)
- micrite (well laminated)
- limestone other than micrite (well laminated)
- massive shell bank
- argillaceous micrite and biomicrite
- thin micrite (well laminated)
- limestone other than micrite (thinly bedded)
- clay and calcareous clay (well laminated)

Vertebrate fossils from Durlston Bay, found by Julian & Vicky Sawyer.
Top: Crocodile vertebra.
Middle: Crocodile femur.
Bottom: Turtle carapace.

Further fossils found at Durlston Bay by Julian & Vicky Sawyer.
Top: Crocodile tooth (*Goniopholis* sp.?).
Above left: Fish scale.
Above right: Crocodile coprolite.
Left: Isolated *Callipurbeckia* fish scale.
Below: A cluster of fish scales.

At the southern end of Durlston Bay lies Durlston Head, which sits on a bed of Portland Stone. Directly above lies the Hard and Soft Cockle Beds, overlain in turn by Lower Building Stones of the Intermarine Beds and the *Corbula* and Beef Beds. The rocks repeat the sequence of those found towards Peveril Point and to the east of the Zig Zag path and these beds are best and more safely examined there.

The section towards Durlston Head is an area of complex faulting but access is difficult, quite dangerous and the cliffs are much overgrown. Access is only safe if taking a route along the foreshore from the Zig Zag path but this necessitates an arduous journey over large boulders and rocks for about half a kilometre.

Southern end of Durlston Bay towards Durlston Head.

SWANAGE BAY

The rocks of Swanage Bay are entirely Cretaceous. They extend from Peveril Point eastwards to Ballard Point and show a section through the Wealden Group, Lower Greensand, Gault, Upper Greensand and Chalk.

WEALDEN GROUP

This excursion begins in the north western part of the bay, where sandstones, mudstones and the Coarse Quartz Grit of the Wealden strata are exposed in the natural cliffs, which are becoming increasingly vegetated. Fossils are not generally common, although occasional dinosaur bone or footprints do occur. Fossils from the Wealden Group can sometimes be washed out of the cliff after heavy rains and end up on the scree slopes. This often represents the best chance of finding fossils from these beds.

Wealden Group displayed in the cliffs at the north-western part of Swanage Bay. Rolled dinosaur bones sometimes show up on the beach here, but identification of the species is usually impossible due to their wave-worn state.

Fragments of lignite (top left & bottom left) formed of fossil plant debris, and naturally compressed peat and wood (top right) are common in these Cretaceous fluvial sediments at Swanage Bay. Photos by Viv Field.

Charcoal found within the plant-debris beds at Swanage Bay indicates floodplain wildfires, which reflect a Mediterranean-type of climate with potentially wet winters with hot, dry summers. Given the rich fossil flora and fauna of the Wessex Formation on the neighbouring Isle of Wight, fossils are generally uncommon in Dorset, however the Wealden exposures at Swanage have yielded fish debris, bones and teeth of crocodiles and evidence of dinosaurs in the past. Rolled dinosaur bones are sometimes found among modern beach deposits within Swanage Bay. At Punfield Cove, in the northern section of the bay, various brackish-water bivalves and gastropods occur in the *Vectis* Formation.

LOWER GREENSAND FORMATION & GAULT

Both the Lower Greensand and Gault are present but generally poorly exposed at Swanage Bay. At the northwest corner, at Punfield Cove, the formations are often obscured by vegetation, particularly the Lower Greensand, or landsliding of the Upper Greensand and Chalk, fallen from Ballard Cliff.

The Gault consists of black clay with some calcareous stone bands. The black clay passes up into loam and then into the Upper Greensand. One of the stone bands might contain the ammonites *Anahoplites picteti* or *Goodhalites delabechei*.

Exogyra bivalve in the Gault of Punfield Cove, Swanage.

Coastal section through the vegetated Lower Greensand and Gault to the Chalk headland below Ballard Down.

UPPER GREENSAND FORMATION

Part of the Upper Greensand Formation is usually well exposed at the base of Ballard Cliff, consisting of green, glauconic sands and a hard blue-grey limestone. Some ammonites are present but in the main, fossils are dominated by bivalves, such as *Amphidonte obliquatum*. Much of the formation is hidden by slipped Chalk but ammonites, such as *Anahoplites*, *Mortoniceras* and *Prohysteroceras* can be found.

Coral (above) and bivalve *Neithea gibbosa* (left) from Swanage Bay.

Upper Greensand section at Swanage Bay, with *Prohysteroceras* and *Mortoniceras* ammonites.

THE CHALK GROUP

The Chalk dominates the cliff section at the far end of Ballard Down, with blocks of Glauconitic Marl, the Cenomanian basement bed strewn across the foreshore beneath the cliff face. The blocks need to be carefully searched, as various ammonites from the Grey Chalk Subgroup are present, including *Acanthoceras* and *Turrilites* from the mid-Cenomanian, *Holaster* echinoids and various brachiopods.

Above but not at beach level, the White Chalk Subgroup is visible in the cliff, so the collector must rely on cliff falls, which are frequent, for collection from boulders found at beach level. As a result, caution is urged. Fallen material is usually with fossils within the blocks from the Lewes Nodular Chalk. Fossil collecting in the fallen blocks is comparatively easy, as there is a high chance of finding various *Micraster* echinoids of many species from either the *Micraster cortestudinarium* or the *Micraster coranguinum* zones, both high in the cliff.

The Chalk can be unusually hard here, so careful use of hammer and chisels will be necessary. Extract enough matrix around the fossils to enable more precise preparation back at home.

The Chalk cliffs of Ballard Down, Swanage Bay, with ammonites *Acanthoceras jukesbrownei* from the Grey Chalk Subgroup (Cenomanian) at this location.

Fossils from the Grey Chalk Subgroup at Swanage Bay.
Top left: *Schloenbachia varians* or *Subprionocyclus normalis*? Top right: Unidentified ammonite.
Centre left: *Manteliceras* sp. Centre right: A *Pecten* bivalve. Bottom left: *Spondylus serratus*.
Found and photographed by Vivien Field.

Enaliornis and *Ichthyornis*, both extinct seabirds of the Late Cretaceous dive and swim dangerously close to a basking *Mososaurus*. Artwork by Andreas Kurpisz.

EXCURSIONS NEAR HENGISTBURY HEAD

Studland Bay & Hengistbury Head

**CHALK GROUP
LONDON CLAY FORMATION
BARTON CLAY FORMATION
BOSCOMBE SANDS FORMATION**

The northern end of Studland Bay.

STUDLAND BAY

The Eocene aged (Lutetian Stage) fossils from the Broadstone Sand Member of the **Poole Formation** (part of the Bracklesham Group) are visible in a small section of Tertiary cliff, to be found at the northern end of Studland Bay. Fossil plants are found in a six-metre bed of grey-sandy coloured clay at Redend Point, just above the Redend Sandstone, but they are generally quite poorly preserved and mostly comprise plant debris, which can be found by splitting any *ex situ* rock. This bed also contains small insects which will need to be obtained by a wet sieving method and the use of a strong hand lens or microscope. Better plant remains of stems and leaves occur in the upper part of the bed, where a darker grey clay crops out (see inset below).

At the southern end of Studland Bay, the White Chalk subgroup from the *Mucronata* Zone of the Campanian stage is exposed. Here, the belemnite *Belemnitella mucronata* might be obtained, along with various bivalves such as the abundant *Inoceramus*, *Mimachlamys mantelliana* (shown below) and *Pseudolimea granulata* along with occasional brachiopods.

The cliffs on the Studland Bay (north) side from Old Harry Rocks. The belemnite *Belemnitella mucronata* is shown.

HENGISTBURY HEAD

The Eocene cliffs at Hengistbury Head are the most eastern section of the Dorset coast. Hengistbury Head is a promontory of low-dipping Eocene sands and clays, with a gravel capping. It is of particular geological interest because of the abundance within the sandy clays of sideritic ironstone nodules. There are rare, hollow moulds of fossil mollusc shells in some of the ironstone nodules, although only a small number of nodules now occur. These are in Middle Eocene strata of about 40 Ma, which were deposited at the junction of estuarine or deltaic conditions. Hengistbury Head provides an excellent opportunity to collect seeds from Barton Age clays. It is the start of an Eocene coastline, which continues into Hampshire, including the highly fossiliferous sites of Barton-on-Sea and Milford-on-Sea. There are no macrofossils here, apart from the occasional mollusc mould. The fossils found here are of microfossil seeds, with a good level of preservation, within a bed of lignite clay from the Boscombe Sands found at the base of the cliff. The bed is easy to find. It is two-metres thick but is not uniform throughout its thickness. The siltiest part of this bed is the most productive for seeds.

Another bed, which is higher in the cliff, is a silty grey bed, sometimes exposed on the shore beneath the cliff, where the sea has washed fallen debris. This bed has a lot of plant material that also includes seeds. The best sections are found where the sea has washed out the middle cliff sections and the basal beds are exposed.

Cliffs at Hengistbury Head.

Eocene strata shown in the cliff face at Hengistbury Head.

The cliffs here are over 35 metres tall, divided into four groups:
Pleistocene River Gravels at the top then followed by:
Warren Hill Sand Member of the Barton Clay Formation. The beds comprise some 10 metres of slightly laminated sands, the upper part being yellow with *Solen* and *Panopea*; the lower part is grey with *Pinna*. The sands are Bartonian in age.
Barton Clay Formation. The beds comprise some three-metres of stiff brown and grey clays, with beds of fine sand. Below are fine grey-brown silty clays with abundant mollusc moulds in nodules and *Nummulites rectus*. Glauconitic fine grey-green clayey sand with some quartz grit and *Nummulites prestwichianus* follows and then coarse-grained sands and quartz grit at the base. The Barton Clay here does not look similar to the Barton Clay found at Barton-on-Sea or Highcliffe and it does not contain obvious and well-preserved shelly fossils, like those at Barton. However, fossils are present. They are entirely from the middle part of the Lutetian Stage of the mid-Eocene but will quite probably elude the majority of geologist visitors to the location.
Boscombe Sand Formation. It is in the final section where fossil seeds can be found. The Boscombe Sands are divided into buff and chocolate coloured sands, with small pebbles, and lignitic sands and clays. It is this bed, which makes up the final two-metres of cliff that is most productive.

The beds at Hengistbury are relatively poor in fossils overall, but the total faunal and floral lists are quite diverse. Bag up samples in heavy duty bags, but make sure you write on them what bed the samples are from and make a note of any important information. The best way to process samples is to wet-wash them, which will also clean the samples. It is not a bad idea to do a second wash to clean them further, as this makes looking at them under a microscope easier. The seeds, unless preserved, can deteriorate very quickly. After a while of being exposed to air, the seeds will open up and 'pop', and once this happens they start to crumble. Therefore, you must preserve them (Paraloid B-72 dissolved in acetone works best) or keep them in airtight containers, with a bag of silica gel.

Looking west beyond Kimmeridge Bay, to Gad Cliff and Brandy Bay.

The final stage of our journey across the east coast of Dorset is now complete. From the Chalk cliffs at Bat's Head, where the excursions began, to the sand and clay cliffs of Hengistbury Head has entailed a journey of just over 48 kilometres, but also a journey through geological time of 157 million years.

This series of books began in the far west of the county, where the 210 million-year-old Triassic rocks of the Penarth Group are encountered at Pinhay Bay and where the UNESCO Jurassic Coast begins. The story of life on Earth unfolds along the Dorset coast like no other place and where the unimaginable diversity and profusion of life forms that occupied our planet are revealed, as fossils, in the rocks.

Opposite page: A selection of micro-fossil seeds found in the Boscombe Sand Formation at Hengistbury Head by Ian Cruickshanks.

ABOUT THE AUTHORS

Steve Snowball worked as a teacher, headteacher and education advisor for 35 years, before retiring to the Dorset coast, to spend time in the area he loves. Turning his fossil collecting enthusiasm into writing books on the subject, this book is the fourth title he has co-written for Siri Scientific Press.

Steve is married with two grown-up children. His other interests include landscape photography, walking the Jurassic Coast with his dogs, painting, pottery and music.

Craig Chivers is a fossil preparator, who is both well established and highly respected in the Dorset area. His skills have enabled him to prepare the *Dracoraptor hanigani* skeleton, the 'Welsh dinosaur,' now exhibited in the Natural History Museum Cardiff.

This book is Craig's third contribution for Siri Scientific Press, as a co-author. Besides collecting fossils, Craig also enjoys boating, scuba diving, drawing and painting.

Steve Snowball & Craig Chivers have also written *A Guide to Fossil Collecting on the West Dorset Coast* (2017) and *A Guide to Fossil Collecting on the South Dorset Coast* (2020), both published by Siri Scientific Press, in which paleo-artist, Andreas Kurpisz, has also contributed. The dinosaur images depicted on these two pages are the work of Andreas Kurpisz.

Andreas Kurpisz lives and works in Berlin. Having completed a degree at polytechnic college, he spent the next five years training in both electronics and mechanotronics.

Andreas is a talented palaeo-artist, who is inspired and influenced by the work of Zdenek Burian and Raul Lunia. Andreas has now collaborated with the authors and contributed his digital palaeo-art and modelling creations on all three Dorset guides published by Siri Scientific Press.

Vivien Field was born and bred on the Dorset Jurassic Coast. She has been collecting fossils since the age of four. Now, having studied Geology at university, she is an Engineering Geologist living and working in Devon. She still regularly goes out collecting, particularly in the Chalk and Greensand. Her favourite fossils to collect are echinoids and fish, although one day she hopes to find a dinosaur … just like any kid, big or small!

BIBLIOGRAPHY

Allison, R. J. *The Coastal Landforms of West Dorset.* London: Geologists' Association, 1992.

Arkell, W. J. *The Jurassic System in Great Britain.* Oxford: The Clarendon Press, 1933.

Arkell, W. J. *The Geology of the Country around Weymouth, Swanage, Corfe & Lulworth.* London: H.M. Stationary Office, 1947.

Cope, J. C. W. *Geologists' Association Guide No 22, Geology of the Dorset Coast.* The Geologists' Association, 2012.

Cox, B. M. *Black Head. Extracted from the Geological Conservation Review, Volume 21: British Upper Jurassic Stratigraphy (Oxfordian to Kimmeridgian) Chapter 2: Upper Jurassic Stratigraphy from Dorset to Oxford.* JNCC, 2001.

Davies, G. M. *The Dorset Coast: A Geological Guide.* London: Thomas Murby & Co., 1935.

Dewey, H. *South West England.* London: H.M. Stationary Office, 1948.

Ensom, P. *Discover Dorset: Geology.* The Dovecote Press, 1988.

Ensom, P. C. & Delair, J. B. *Dinosaur Tracks from the Lower Purbeck Strata of Portland, Dorset.* Proceedings of the Usher Society, 2007.

House, M. *The Geologists' Association Guide No 22, Geology of the Dorset Coast.* The Geologists' Association, 1993.

Lomax, D. R. & Tamura, N. *Dinosaurs of the British Isles.* Siri Scientific Press. Manchester, 2014.

Milner, A. R. *The Turtles of the Purbeck Limestone Group of Dorset, Southern England. Palaeontology, Vol. 47, Part 6.* The Palaeontological Association, 2004.

Natural History Museum. *British Mesozoic Fossils.* London: H.M. Stationary Office, 2013.

Scriven, S. *Fossils of the Jurassic Coast.* Jurassic Coast Trust, 2016.

Smith, A. B. & Batten, D. J. Fossils of the Chalk. London: Palaeontological Association, 2002.

Sweetman, S.C., Smith, G. & Martill, D.M. Highly derived eutherian mammals from the earliest Cretaceous of southern Britain. Acta Palaeontologica Polonica 62 (4): 657–665, 2017.

Swinton, W. E. *Fossil Amphibians and Reptiles*. London: Printed by order of the Trustees of the British Museum (Natural History), 1962.

Westacott, H. D. *The Dorset Coast Path*. Penguin, 1982.

Wilson, V., Welch, F. B. A., Robbie, J. A. & Green, G. W. *The Geology of the Country around Bridport & Yeovil*. London: H.M. Stationary Office, 1958.

Wright, J. K. *Osmington*. Extracted from the Geological Conservation Review Volume 21: British Upper Jurassic Stratigraphy (Oxfordian to Kimmeridgian) Chapter 2: Upper Jurassic stratigraphy from Dorset to Oxford. JNCC, 2001.

Wright, J. K. *East Fleet*. Extracted from the Geological Conservation Review Volume 21: British Upper Jurassic Stratigraphy (Oxfordian to Kimmeridgian) Chapter 2: Upper Jurassic stratigraphy from Dorset to Oxford. JNCC, 2001.

WEBSITES

http://www.soton.ac.uk/~imw/index.htm West, Ian M. 2019. Geology of the Wessex Coast .

https://ukfossils.co.uk Cruickshanks, Alister. UK Fossils Network.

http://www.discoveringfossils.co.uk/ Shepherd, Roy.

https://www.siriscientificpress.co.uk

INDEX OF FOSSIL PHOTOGRAPHS

A
Acanthoceras jukesbrownei 26, 175
Aequipecten aspera 84
Archaeoniscus broiei 138

B
Belemnitella mucronata 181
Brachiosaurus sp. 133, 136

C
Callihoplites sp. 98
Callipurbeckia minor 36, 137, 143, 152, 169
Catrurus sp. 110, 120
Calycoceras sp. 86
Calycoceras gentoni 100
Calycoceras newboldi 100
Catopygus carinatus 93
Coelodus sp. 152
Corbula alata 160
Cuttlefish 28

E
Epivirgatites sp. 87
Eurycormus sp. 120
Exogyra sp. 173

G
Glaucolithites sp. 87
Goniopholis simus 154, 156

H
Hylaeochelys sp. 90

I
Ichthyosaurus sp. 108, 109
Iguanodon sp. 143, 159
Inoceramus sp. 80

L
Lepidotes sp 24

M
Macromesodon daviesi 30, 138
Mantelliceras sp. 98
Micraster coranguinium 97
Micraster cortestudinarium 82, 97
Mortoniceras sp. 174

N
Nanogyra nana 88
Neomiodon medius 160
Neithea gibbosa 174

O
Osteichthyes sp. 108
Ostlingoceras sp. 81

P
Pectinatites sp. 113
Pachythrissops sp. 121
Plant debris 172
Pliosaurus brachyspondylus 115
Praeexogyra distorta 96, 153
Prohysteroceras sp. 174
Ptychodus sp. 85

S
Salenia petalifera 93
Sauropod 158
Scapanorhynchus sp. 85
Scaphites aequalis 100
Scaphites obliquus 100
Schloenbachia sp. 86
Sternotaxis planus 80
Stoliczkaia sp. 98
Spondylus serratus 176

T
Theropod 158
Thrissops sp. 121
Titanites anguiformis 126, 128

Jurassic Coast Guides

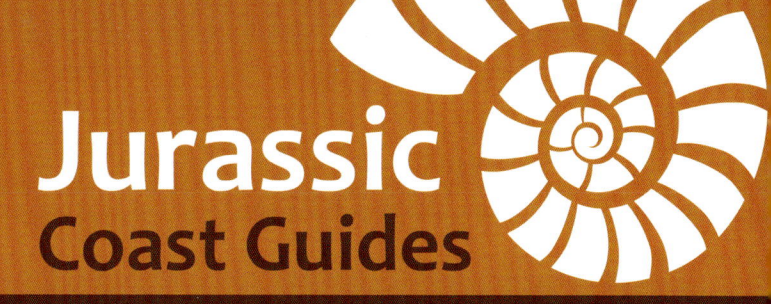

Professionally guided walks, tours and fossil hunts

Private Fossil Walks last for 3 hours and are held at Charmouth, Dorset. All walks are dependent on tide times and the weather conditions. Walks start with a safety talk and a discussion on how to find fossils on the foreshore at Charmouth. Examples of fossils that can be found are shown before heading down onto the beach to hunt. Great fun and enjoyable for all ages.

Group Fossil Walks are available to book during the school holidays.

'Guided Tour Of The Year!'
GOLD AWARD - Dorset Tourism Award 2018

Dorset Tourism Awards 2018 GOLD AWARD
Dorset Tourism Awards 2019 GOLD AWARD

Corporate Events

Jurassic Coast Tours

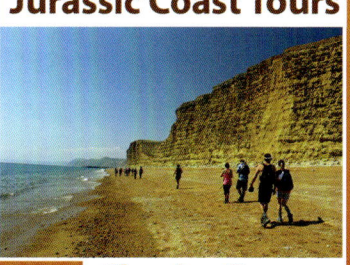

Charity Events

2020 Travellers' Choice - Tripadvisor

Guided walks, tours and fossil hunts can be tailored to suit all ages and abilities. They are led by qualified, insured & experienced Mountain Leaders. They can be arranged for individuals, couples, families or groups. We have over 10 years of experience guiding walks along the Jurassic Coast and the surrounding areas.

We also arrange guided charity walking challenges, Corporate walking events, VIP tours and Jurassic Coast tours from Portland Port.

+44 (0)7900 257944
guide@jurassiccoastguides.co.uk
www.jurassiccoastguides.co.uk

© 2021 Jurassic Coast Guides

Natural Selection Fossils
est. 2004

www.naturalselectionfossils.com

Natural Selection Fossils was established in 2004 selling fossils online. We specialise in preparing fossils to a high standard and also creating detailed resin replicas of various specimens.